KEY CONCEPTS IN POLITICAL GEOGRAPHY

KEY CONCEPTS
IN HUMAN GEOGRAPHY

The *Key Concepts in Human Geography* series is intended to provide a set of companion texts for the core fields of the discipline. To date, students and academics have been relatively poorly served with regards to detailed discussions of the key *concepts* that geographers use to think about and understand the world. Dictionary entries are usually terse and restricted in their depth of explanation. Student textbooks tend to provide broad overviews of particular topics or the philosophy of Human Geography, but rarely provide a detailed overview of particular concepts, their premises, development over time and empirical use. Research monographs most often focus on particular issues and a limited number of concepts at a very advanced level, so do not offer an expansive and accessible overview of the variety of concepts in use within a subdiscipline.

The *Key Concepts in Human Geography* series seeks to fill this gap, providing detailed description and discussion of the concepts that are at the heart of theoretical and empirical research in contemporary Human Geography. Each book consists of an introductory chapter that outlines the major conceptual developments over time along with approximately twenty-five entries on the core concepts that constitute the theoretical toolkit of geographers working within a specific subdiscipline. Each entry provides a detailed explanation of the concept, outlining contested definitions and approaches, the evolution of how the concept has been used to understand particular geographic phenomenon, and suggested further reading. In so doing, each book constitutes an invaluable companion guide to geographers grappling with how to research, understand and explain the world we inhabit.

Rob Kitchin
Series Editor

KEY CONCEPTS IN POLITICAL GEOGRAPHY

CAROLYN GALLAHER
CARL T. DAHLMAN
MARY GILMARTIN
AND
ALISON MOUNTZ
WITH PETER SHIRLOW

SAGE

Los Angeles | London | New Delhi
Singapore | Washington DC

First published 2009

SAGE Publications Ltd
1 Oliver's Yard
55 City Road
London EC1Y 1SP

SAGE Publications Inc.
2455 Teller Road
Thousand Oaks, California 91320

SAGE Publications India Pvt Ltd
B 1/I 1 Mohan Cooperative Industrial Area
Mathura Road, Post Bag 7
New Delhi 110 044

SAGE Publications Asia-Pacific Pte Ltd
33 Pekin Street #02-01
Far East Square
Singapore 048763

Library of Congress Control Number: 2008938831

British Library Cataloguing in Publication data

A catalogue record for this book is available from the British Library

ISBN 978-1-4129-4671-1
ISBN 978-1-4129-4672-8 (pbk)

Typeset by C&M Digitals (P) Ltd, Chennai, India
Printed in Great Britain by CPI Antony Rowe, Chippenham, Wiltshire
Printed on paper from sustainable resources

Mixed Sources
Product group from well-managed forests and other controlled sources
www.fsc.org Cert no. SGS-COC-2953
© 1996 Forest Stewardship Council
FSC

CONTENTS

ACKNOWLEDGEMENTS

Figure 1.1: A fortified portion of the border between Mexico and the US
Source: IStock © Paul Erickson

Figure 1.2: Map of Basque Provinces in Spain and France
Source: http://commons.wikimedia.org/wiki/Atlas_of_the_Basque_Country

Figure 1.4: Voortrekker Monument, Pretoria, South Africa
Source: IStock © Hansjoerg

Figure 3.1: Demonstrators march through the intersection of 18th and M Streets NW in Washington DC at a demonstration against the World Bank and IMF 16 April 2005
Source: http://commons.wikimedia.org/wiki/File:A16_IMF_march.jpg
Photo by Ben Schumin

Figure 3.2: Mainstream march, part of the October Rebellion demonstrations against the World Bank and IMF in Washington DC
Source: http://commons.wikimedia.org/wiki/File: October_Rebellion_mainstream_march_3.jpg
Photo by Ben Schumin

Figure 4.3: Zapatista Primary School
Source: Everystock

Figure 5.1: Antonio Gramsci, around 30 years old, c. 1920
Source: http://commons.wikimedia.org/wiki/Image:Gramsci.png

Figure 5.2: Remnants of the Berlin Wall
Source: IStock © Carsten Madsen

Figure 8.4. Chinese nuclear weapons can now reach targets in most of the world except South America (Courtesy Office of the Secretary of Defense)
Source: Office of the Secretary of Defense. (2007). *Annual Report to Congress: Military Power of the People's Republic of China 2007* Washington, D.C.: Department of Defense, p. 19

Figure 9.1: Halford Mackinder
Source: http://commons.wikimedia.org/wiki/Category:Mount_Kenya

Figure 9.3: Elisée Réclus
Source: http://commons.wikimedia.org/wiki/Image:EliseeReclusNadar.jpg

Figure 10.1: Drawing of Adam Smith
Source: http://commons.wikimedia.org/wiki/Image:AdamSmith.jpg

Figure 10.2: Karl Marx 1866
Source: http://commons.wikimedia.org/wiki/Image:Marx1866.jpg

Figure 10.3: The dungeon at Elmina Castle, Ghana housed slaves bound for Brazil and the Caribbean
Source: IStock © Alan Tobey

Figure 10.5: Robert Clive, the British governor of India is seen receiving a decree from Shah Alam, the Mughal Emperor of India, guaranteeing the East India Company the right to administer the revenues of Bengal, Behar and Orissa
Source: IStock

Figure 11.1: Remains of the Berlin Wall, which divided East and West Germany
Source: IStock © Georg Matzat

Figure 11.2: The Vietnam War Memorial in Washington DC. Vietnam was a site of one such conflict
Source: IStock © Robert Dodge

Figure 11.3: Istanbul University
Source: IStock © Simon Podgorsek

Figure 13.2: Bolivia's major cities (Map courtesy of the University of Texas Perry-Castañeda Library map collection)
Source: University of Texas Perry-Castañeda Library map collection
http://www.lib.utexas.edu/maps/cia07/bolivia_sm_2007.gif

Figure 13.4: Riot police in the streets of La Paz, Bolivia
Source: IStock

Figure 14.1: Georgian President Mikheil Saakashvili at the opening ceremony for a Pepsi bottling plant
Source: http://commons.wikimedia.org/wiki/Image:Mikheil_Saakashvili_at_softdrink_factory_2004-May-25.jpg

Figure 14.2: McDonald's restaurant in Seoul, Korea
Source: http://commons.wikimedia.org/wiki/Image:Mcdonalds_seoul.JPG

Figure 14.4: Oaxaca is one of Mexico's poorest states
Source: http://commons.wikimedia.org/wiki/Image:Oaxaca_en_mexico.svg

Figure 15.1. The border crossing between Tijuana, Mexico, and San Diego, California in the US
Source: IStock © James Steidl

Figure 15.2: Electric fence at Baka-El-Garbia separating Israel from Palestine
Source: IStock © Dor Jordan

Figure 15.3: Refugee Camp in Northern Sudan
Source: IStock © Claudia Dewald

Figure 17.1: Fence topped with barbed wire in the demilitarized zone between North and South Korea
Source: IStock © Eugenia Kim

Figure 17.2: Concrete wall separating Israel from Palestine Bakka-El-Garbia, Israel
Source: IStock © Dor Jordan

Figure 17.3: The front entrance to a gated community in the US
Source: IStock © Rich Legg

Figure 17.4: Australia's 'Pacific Solution', Christmas Island, Nauru and Melvile Island
Source: Joe Stoll and Syracuse University Cartography Laboratory

Figure 19.4: Narmada Dam (Sardar Samovar Dam), India
Source: Everystock

Figure 21.1: A Cuban stamp commemorating Che Guevara
Source: IStock

Figure 21.2: Propaganda poster with image of Osama bin Laden on it
Source: Wikicommons http://en.wikipedia.org/wiki/Osama_bin_Laden

Figure 21.3: Map of Israel with major cities
Source: University of Texas Perry-Castañeda Library Map Collection http://www.lib.utexas.edu/maps/cia07/israel_sm_2007.gif

Figure 21.4: Map of the West Bank with major cities
Source: University of Texas Perry-Castañeda Library Map Collection http://www.lib.utexas.edu/maps/cia07/west_bank_sm_2007.gif

Figure 22.1: Pyotr Kropotkin
Source: http://commons.wikimedia.org/wiki/Image:Kropotkin_Nadar.jpg#file

Figure 22.2: In the mid to late 1980s the city of Baltimore, Maryland, adopted a festival marketplace, turning the city's old shipyards into "Inner Harbor"– a tourist and entertainment centre
Source: IStock © Gabriel Eckert

Figure 22.3: The high rise, luxury condominium tower Harborview stayed afloat with a $2 million tax relief plan from the city of Baltimore
Source: IStock © Gabriel Eckert

Figure 23.1: The Quebec flag painted on a young boy's face
Source: IStock © Duncan Walker

Figure 23.2: Map of Sri Lanka: shaded area is claimed by the Tamil Tigers (as of August 2005)
Source: http://commons.wikimedia.org/wiki/Image:Location_Tamil_Eelam_territorial_claim.png

Figure 23.3: Russian helicopter downed during the First Chechen War
Source: http://upload.wikimedia.org/wikipedia/commons/6/6a/Evstafiev-helicopter-shot-down.jpg

Figure 23.4: Chechen Fighter during the Battle for Grozny
Source: http://commons.wikimedia.org/wiki/Image:Evstafiev-chechnya-tank-helmet.jpg

Figure 24.1: California Driver's license
Source: IStock © Hal Bergman

Figure 24.2: Concentration camp overalls with a Star of David patch
Source: IStock © Stephen Mulcahey

Figure 24.3: Estonia within EU countries as of 2007
Source: http://upload.wikimedia.org/wikipedia/commons/0/0c/EU_location_EST.png

Figure 25.1: General Post Office, Dublin, Ireland, a contested site in Irish colonial and postcolonial history
Source: IStock © Mark Joyce

Figure 25.2: Edward Said
Source: http://commons.wikimedia.org/wiki/Image:Edward-Said.jpg

Figure 25.3: Map of Afghanistan-Pakistan border
Source: University of Texas Perry-Castañeda Library map collection. http://www.lib.utexas.edu/maps/middle_east_and_asia/afghan_paki_border_rel88.jpg

Figure 27.3: Symbol for transgender/intersexed people
Source: http://commons.wikimedia.org/wiki/Image:A_TransGender-Symbol_Plain3.svg

Figure 27.4: Photo collage of disappeared people during Argentina's dirty war
Source: http://upload.wikimedia.org/wikipedia/commons/e/ed/Que_digan_d%C3% B3nde_estan.jpg

Figure 28.1: An illustration in the *London News* of Orientalists inspecting the Rosetta Stone during the International Congress of Orientalists, 1874
Source: http://upload.wikimedia.org/wikipedia/commons/5/57/Rosetta_Stone_International_Congress_of_Orientalists_ILN_1874.jpg

Figure 28.2: *The Grand Odalisque*, by Jean Auguste Dominique Ingres, 1814
Source: http://upload.wikimedia.org/wikipedia/commons/d/d8/Jean_Auguste_Dominique_Ingres_005.jpg

Figure 28.3: Gay pride parade in Toronto, Canada, 2005
Source: IStock © Arpad Benedek

INTRODUCTION

The world of politics provides plenty to whet the appetite. The anticipation of an upcoming election, the intrigue behind a *coup d'état*, and the chaos of a war zone are gripping fare for the intellectual, student and layperson alike. What unites geographers in their study of political events like these is a focus on the spatial organization inherent to them and the power relationships that underpin them. Geographers look at how politics affect spatial order, and how spatial orders inform politics.

Traditionally, political geography has used the state as a primary unit of analysis. Political geographers studied how states were organized internally, and how they interacted with other states in regions and the international system as a whole. In recent years political georaphy has added other units of analysis to its repertoire. These include not only smaller levels of analysis, such as the 'local', but also larger ones, such as the supranational. Their use has also brought renewed attention to the different ways that political actions play out *across* scales.

In many ways this change in focus reflects changes in the world around us. When the Cold War ended in 1989 there was uncertainty not only about what would happen to formerly communist states, but also what would happen to the balance of power between states. The emergence of globalization also brought new political actors to the fore, including international organizations, social movements, non-governmental organizations and warlords, among many others. How this mix of old and new actors and the changing relations of power between them will play out is yet to be seen, but political geography will be there to document, analyse and ultimately theorize them.

Political Geography through Time

The development of modern political geography was intimately connected with the colonial project (Peet 1985). These connections are readily apparent in the subdiscipline's two most formative schools of thought – environmental determinism and geopolitics. While these approaches initially made the discipline of Geography popular in and out of the academy, they would eventually be debunked, leaving political geography fighting for its survival. A brief introduction to each is provided here. Geopolitics, which has witnessed a resurgence of interest under the label 'critical geopolitics', is also discussed in Chapter 7.

Environmental Determinism

Environmental determinism was developed in the mid-nineteenth century purportedly to explain the discrepancies in standards of living between European colonizers and their colonial subjects. Environmental determinists were influenced by social Darwinism, although most preferred to draw from Lamarckian rather than Darwinian versions of evolution (Livingstone 1992).[1] Proponents of the theory, including Friedrich Ratzel, Ellen Churchill Semple and Ellsworth Huntington, posited that climate and topography determined the relative development of a society, and its prospects for future development. Temperate climates were seen as invigorating whereas tropical and arctic climates were deemed to stunt human development. Geographers also postulated that river valleys produced vibrant societies while mountainous environments inhibited them.

For much of the early twentieth century, especially in the United States, environmental determinism dominated the entire discipline. Even as the approach was becoming a meta-narrative of the field, scholars in other disciplines were subjecting it to withering criticism. The anthropologist Franz Boas labelled the theory simplistic and reductionist because it failed to explain how vastly different cultures could emerge in the same environments (Livingstone 1992). Eventually, geographers would abandon the theory as well. One of the first to do so was Carl Sauer who adopted culture, rather than environment (alone), as the key explanatory variable in human differentiation across space (Livingstone 1992). Half a century later, geographers would describe the discipline's fixation with geographic determinism as an imperialist impulse (Peet 1985; Smith 1987).

Geopolitics

Social Darwinism also influenced political geographers' view of the state. Most notable in this regard was Friedrich Ratzel, whose book *Anthropogeography* formed the basis for environmental determinism.

Ratzel theorized that states were very much like organisms; both had life cycles, and states, when they were young, needed *lebensraum*, or living space, to grow. Ratzel's theory of *lebensraum* was further developed in the German context by Karl Haushofer and Richard Hennig and in Britain by Halford MacKinder (Livingstone 1992). In the 1930s, Nazi ideology combined the geopolitical view of state life cycles, and the territorial imperative underpinning them, with eugenics (Livingstone 1992). After the Nazi atrocities were brought to light at the end of World War II, geopolitics looked as ill-conceived as environmental determinism had before it. As disciplinary historian David Livingstone (1992: 253) succinctly observes, these schools of thought failed to separate out 'the science of geography from practical politics'.

Theoretical Influences

In many respects, political geography is an empirically driven sub-discipline (Mamadouh 2003). Political geographers tend to employ mid-level concepts rather than meta-theories to analyse the spatial organization of politics. Historically, concepts like region, territory and scale gave the sub-discipline its coherence, with debates emerging around how these concepts should be defined and employed. The focus on regional studies during the Cold War buttressed this trend in Anglo-geography as political geographers worked to build a dossier of thick, in-depth knowledge on places deemed of political importance by the government and/or military establishment. When political geographers use meta-level theory, they tend to select from two general theoretical frames: political economy and poststructuralism.

Political Economy

Although some of the discipline's key thinkers can reasonably be labeled Marxist geographers, most political geographers borrow from the Marxist canon rather than working fully within it. These approaches are generally termed political economy to indicate that economic structures are emphasized in the analysis of the political realm. Several schools of thought can be broadly fitted under the political economy framework. These are discussed below, although the reader should refer to Chapter 10 on Political Economy for a more detailed account of the genesis of the term and its uses in political geography.

World Systems Theory

World Systems Theory (WST) posits that macro-level patterns govern social and economic change. Although popular in geography, WST was developed by a political scientist, Immanuel Wallerstein (1974). Wallerstein wanted to challenge conventional notions of economic development in both history and the social sciences. Drawing on the work of French historian Fernand Braudel (1993), Wallerstein argued that history was not about singular events – the start of a war or the signing of a diplomatic accord – but about materially structured ways of life (Wieviorka 2005). Understanding history required understanding the material foundations of society, not just the actions of its elites. Wallerstein's work was also informed by the work of Andre Gunder Frank (1969), who criticized contemporary understandings of modernization, or a lack thereof, in the developing world (Wallerstein 2006). Frank argued that countries did not develop (or fail to develop) simply because they had taken (or failed to take) the necessary steps; they developed (or failed to develop) because of their place in the colonial order. Wallerstein posited that the contemporary world system emerged during the colonial period and was consolidated as a world (rather than a regional) system by the early 1900s. This process created a spatial structure in which so-called core countries were able to develop economically and politically through the extraction of peripheral countries' surplus.

Geographers who use WST tend to employ the approach in one of two ways (Flint and Shelley 1990). In the first, they use the model largely as it is, or with minor variations, to frame their analysis of political and economic change in given states or regions. Jim Blaut (2000) has used the theory, for example, to criticize historians who argue that the industrial revolution happened in Europe rather than Africa, Latin America or Asia because it possessed special features these other regions did not. Rather, the extraction of peripheral countries' surplus allowed core countries to develop, at the expense of those places whose surplus they took. In other cases, geographers have nuanced the model, examining how world systems theory can be applied at different scales. They note that the categories of core and periphery are better seen not as static places on the globe but as scalar process (Dyke 1988;

Straussfogel 1997). So conceived, apparent contradictions in the model (such as the appearance of peripheral-like places in the core – e.g. Appalachia in the US) can be explained. That is, core/periphery relationships operate not only at the global scale, but also at national and even local scales.

Regulation theory

The regulation school was developed in France by Michel Aglietta (1979) and Alain Lipietz (1992) in the late 1970s and 1980s. Regulation theory is not Marxist, *per se*, but its advocates accept the Marxist notion that the capitalist system is prone to crisis (Purcell and Nevins 2005). In particular, they argue that capitalism is subject to crises of accumulation and these will eventually lead to the collapse of the entire system. Capitalism has, of course, proven to be quite durable, and the regulation school developed to explain what has kept its collapse at bay.

Regulation theory was developed in the late 1970s at a time of crisis. The OPEC oil embargo in 1973 had flooded the world's financial system with petrodollars (the profits gained from a reduced supply and increased prices). Many of these dollars were invested in western banks, which redistributed them as loans to domestic and foreign borrowers. In the US, this extra cash helped contribute to stagflation, and led Paul Volcker, then chairman of the Federal Reserve Board, to enact sharp interest rate hikes beginning in 1979. Access to credit dried up and unemployment increased in the US and its trading partners. Meanwhile, in the developing world, many countries who were the recipients of loans financed by petrodollars saw their debt skyrocket under the higher interest rates. By 1982 several of the world's countries were on the verge of default. Regulation theorists sought to understand the crisis of the late 1970s by examining what states had done in prior periods to cultivate relative stability in the system; what Lipietz (1992) labels the 'grand compromise' between the state, capital and labour.

Geographers in a variety of sub-disciplines have employed regulation theory (Smith and Pickles 1998). For their part, political geographers have tended to use regulation theory to examine how states manage their economies in order to avoid a crisis of accumulation (Jessop 1995; Jones 2001; Purcell and Nevins 2005). Regulation theory's emphasis on managing competing class interests has also given rise to studies examining how states manufacture the consent of their populations for changes that may be unpopular, such as raising interest rates or increasing taxes (Jessop 1997; Purcell and Nevins 2005). Political geographers have also examined how labour can win concessions in a mode of production less friendly to them than Fordism (Herod 2000).

Political ecology

Political ecology allows geographers to examine how the physical environment and processes affecting it, such as deforestation or climate change, are connected to human activity, generally, and societal modes of production more specifically (Blaikie and Brookfield 1987; Peluso and Watts 2001; Robbins 2004). A central premise of political ecology is that no ecology is 'apolitical', even though we often assume the contrary. Driving through Yellowstone National Park, for example, political ecologist Paul Robbins (2004: xv) observed that the park's presumably wild terrain has been subject to human imperatives for millennia. As he explains, Native American hunting patterns 'probably served to concentrate the elk, antelope and other animals that made the site so attractive to Anglo-Americans who later occupied the land'. Likewise, the near extinction of wolves by westward development prompted booming elk populations in the area. Park managers responded by culling the herds, which triggered protests from those who thought natural predators should do the job instead, leading eventually to the reintroduction of wolves to the park. In short, Yellowstone's 'wild' landscape is the product of all sorts of human decisions, which were themselves the product of political institutions.

Political ecology addresses four areas of concern: land degradation, environmental conflict, conservation efforts, and more recently environmental social movements (Robbins 2004). In all of these areas, political ecologists tend to go against the grain. For example, while mainstream analysis of land degradation places blame on the poor land management techniques used by peasants, political ecology points to state policies which often force peasants to use land more intensively in order to meet their basic subsistence needs.

Poststructuralism

Poststructuralism is a theoretical perspective that emphasizes language and the production of meaning in the analysis of societal relations. The emergence of poststructuralism in the social sciences is often referred to as the 'linguistic turn' and is associated with French scholars who came of age in the 1960s, including but not limited to Jacques Derrida, Michel Foucault, Gilles Deleuze, Félix Guattari and Chantal Mouffe (Howarth 2000).

The use of the prefix 'post' to describe the theoretical frame developed by these scholars is a bit misleading. Poststructuralists did not abandon structure so much as change their notion of it. In the social sciences poststructuralists often self-consciously pitted themselves against the Marxist tradition (Bondi 1993). They argued that economic structure could not adequately capture the human experience. While one's class positioning could explain some facets of exploitation, for example, it failed to take account of abuses carried out on the basis of gender, race, sexuality or national origin. As such, poststructuralists examined how the social categorization of dominant and weaker groups was normalized and rationalized through language. This focus is often referred to broadly as identity politics because the study of dominance and 'otherness' often boils down to how people are defined in society and how they maintain, resist, subvert or nuance those identities.

In this way poststructuralism also represents a critique of wider social science epistemology. That is, while traditional social science disciplines hold that the production of knowledge is neutral and objective, poststructuralists believe that knowledge production is political and so all truth claims are constitutive of the political orders of which they are a part. When leaders assert, for example, that a particular alliance is necessary or that a war is inevitable, poststructuralists deconstruct these claims rather than take them at face value. Deconstruction is a common method in poststructural analysis; it attempts to examine why truth claims are created and how they are naturalized. In geography poststructuralism has manifested in one of two broad ways – in feminist approaches and in critical geopolitics.

Feminism(s)

Some of the earliest forays into poststructuralism in geography broadly, and political geography more specifically, have come from feminists (Bondi 1990, 1993; Sharp 1996). The emergence of feminism in geography was both a political and an analytic venture. Politically, early feminists argued against the exclusion of women as geographic topics of study. In a now seminal piece in the *Professional Geographer*, for example, Janice Monk and Susan Hanson (1982) observed that while the discipline purported to describe and analyse the spatial patterns of humans it was actually doing so only for the male half of the population. Over time, feminists turned to analytic concerns as well. In particular, they argued that societal relations were gendered. Societies from North America to sub-Saharan Africa assume certain roles for women and men, and these broadly accepted assumptions shape what people do in life and how they are regarded, especially when they step outside of expected (and accepted) gender norms.

Within political geography feminism has focused on a variety of themes. A number of feminist political geographers have contributed to postcolonial studies, a cross-disciplinary topic that deals with issues in postcolonial societies and borrows heavily from poststructural theories of difference (Pratt 2000). Relatedly, feminists in political geography have examined how 'otherness' – social categories outside normative identity constructions – is spatialized. Gil Valentine's work on lesbian geographies (1994), for example, notes that lesbians have had more difficulty than gay men in creating specifically lesbian spaces in cities so they have tended to focus on creating more informal and mobile gathering places: events at clubs, friends' houses, etc. Feminists have also worked to 'gender' classic concepts in political geography, such as nationhood (Sharp 1996). More recent interventions include Hyndman's call for a feminist geopolitics (Hyndman 2001, 2004a).

Critical geopolitics

Critical geopolitics is another avenue of exploration within political geography that has been influenced by poststructuralism (Ó Tuathail and Dalby 1998). Critical geopolitics is an attempt to 'radicalize' geopolitics. It rejects the traditional understanding of geopolitics as 'a neutral and objective practice of surveying global space' (Ó Tuathail and Dalby 1998: 2). Instead, critical geopolitics holds that all truth claims are political: that they are made on behalf of vested political interests and often in the pursuit of political economic imperatives. In this way critical geopolitics manifests a classic concern of poststructuralism to highlight the contingent and political nature of knowledge production.

While there is no thematic 'centre' to critical geopolitics, scholars working within the approach tend to focus on unpacking geopolitical claims. This has, by necessity, led to a concentration on the production of discourse. In poststructural theory discourse is more than rhetoric. It is a linguistic structure of meaning through which social, economic and political hierarchies are established and then legitimized. A number of studies of the colonial period note, for example, that the discipline of geography helped justify colonialism by invoking neo-Lamarckian discourses on racial difference (Driver 1999; Godlewska and Smith 1994; Kearns 1993). More recent studies have examined how the imposition of free market policies in the West and in developing countries was designed to benefit financial interests over and against those of producers (Harvey 2000; Ould-Mey 1996). More information on critical geopolitics and its application to studies of colonialism and free market 'reforms' can be found in Chapters 7, 9 and 10 (on Geopolitics, Colonialism/imperialism and Political Economy, respectively).

Fault Lines in a Subdiscipline

Most disciplines contain intellectual and political fault lines. Geography is no exception. Although students can get a Bachelor, Master's, or doctoral degree in Geography, most students follow either a human or a physical track as they fulfil major requirements. And, once students have chosen a track, they are often expected to specialize in a subfield. At graduation, a human geography student with a focus on political geography may know very little about physical geography or even other human geography subfields. Perhaps not surprisingly, sub-disciplines behave in a similar fashion to the wider disciplines whence they stem. In political geography, there are a number of internal fault lines. Two of the most trenchant are discussed here.

The Regional versus the Thematic

While all political geographers want to know the world, they often disagree on how best to know it. One of the subdiscipline's longest-standing debates has been between those who think the pursuit of geographic knowledge should be based in regions and those who think it should be thematically organized.

In the 1950s the debate came to a head when Fred Schaefer (1953) published an article in the *Annals of the Association of American Geographers* that rejected the then prevailing regional approach in the discipline. He argued that geographers should adopt a more scientific approach to the discipline; they should delineate the key spatial patterns associated with human behaviour and uncover the 'rules' or 'laws' that underpin them. Richard Hartshorne, who had done much to put regional geography at the core of the discipline (see Hartshorne 1939) in the decades prior to Schaefer's article, responded by vigorously defending the need for a regional-based curriculum (1954, 1955). Within a decade of the debate, however, Schaeffer's approach was in the ascendant. Indeed, the growing sway of positivism in the social sciences during the 1960s and 1970s favoured a systematic rather than place-specific approach.

The debate has continued, albeit periodically, in the years since. The regionalist viewpoint, for example, re-emerged in the early 1980s when John Fraser Hart wrote an article for the *Annals of the Association of American Geographers* describing regional geography as 'the highest form of the geographer's art' (1982). Like Schaeffer's challenge almost thirty years earlier, Hart's engendered swift and vigorous defences of the systematic approach (Golledge et al. 1982; Healey 1983). However, others in the discipline responded that the debate was 'sterile' and not particularly relevant (Smith 1987). Likewise, Mary Beth Pudup (1988: 385) argued that the debate missed a wider point – that regional geographers need a 'theory of description' to guide their 'interpretive quest'.

After the 9/11, 2001 attacks on the World Trade Center in New York and the Pentagon in Washington, DC, proponents of the regional approach re-emerged (Toal 2003; Wade 2006). Gerard Toal (2003), for example, made a strong case for 're-asserting the regional' in political geography. He argued that the American response to the attacks represented a 'clash of ignorance'. That is, George W. Bush and Osama bin Laden held stereotypical and messianic views of one another. In the case of George Bush, the result was a simplistic geopolitics that divided the world between the 'free' and the 'evil'. Situations on the ground, in Afghanistan and Iraq, are of course much messier than such a dichotomy suggests. And that simplistic and messianic vision that underpins the so-called 'war on terror' was the Achilles' heel of Bush's foreign policy. Toal argued that only a thick, regional knowledge can help the United States come to grips with the threat posed by Al Qaeda, specifically, and with the US's changing role in the post-Cold War epoch more broadly.

While the debate between the regional and thematic approaches in geography is likely to continue, it is worth noting that many people in the discipline work every day to merge, blend or use both approaches (see Steinberg et al. 2002 for a good overview).

Politics versus Politics

For much of political geography's history, the politics under consideration was of the 'big P' variety. 'Big P' politics has traditionally dealt with states and their relations with other states or groups of states. Geopolitics, for example, concerned itself with the way that states manipulate territory to their advantage. During the 1980s and 1990s, the influence of poststructual forms of analysis, especially feminism, ushered in a new focus on so-called 'small p' politics. 'Small p' politics includes politics by non-state actors who tend to work through social movements and other collectives rather than political parties and other state-centred institutions.

The divide between small and big 'p' forms of analysis has played out in a variety of ways. Demographically, the divide is often, though not exclusively, generational. Young scholars often cut their teeth on what Colin Flint calls 'post-1960 [political] issues' – identity politics built around gender, sexuality, race, ethnicity and the environment. However, because mainline political geography has traditionally been focused on 'big P' forms of analysis, many people doing 'small p' studies do not describe themselves as political geographers. As Flint (2003a: 618) notes, most scholars doing 'small p' work are 'not 'card-carrying political geographers' even though they are doing political geography.

The social distance between 'small p' and 'big P' studies is particularly evident among the feminists working within the political geography tradition. While a feminist political geography emerged as early as the mid-1980s, the majority of the subdiscipline has ignored, or given only scant attention to, its findings in the years since its arrival (Hyndman 2004a; Sharp 2007). As Sharp observes, of the discipline's subfields, political geography 'has been least influenced by feminist approaches and least inclusive of female geographers' (2007: 382). Feminist political geographers have explained this state of affairs in a variety of related ways. At a general level, many observe that the subfield is dominated by men and as such reflects the interests and biases of those who dominate it (Sharp 2000). At an epistemological level, Staeheli and Kofman (2004) argue that political geography is 'masculinist'. That is, the subdiscipline refuses to include gender as an important variable of political analysis, and by doing so, assumes, incorrectly, that political change only results from the actions of (male) elites. In a similar vein, Sharp (2000) argues that political geography's reluctance to move beyond the study of statecraft leads it to ignore how political change is embodied in everyday, local practices. Examining these processes is important because such practices often contradict formal discourses about the way political change is said to occur and can complicate our understanding of the reasons for, and the effects of, traditional statecraft.

As several scholars have noted, few political geographers have answered the feminist clarion call in the discipline (for an exception see Painter 1995). When they have addressed feminist concerns, they have often done so in a way that suggests that 'small p' concerns are not really the purview of the subdiscipline. In a 2003 forum on the state of political geography, for example, John Agnew argued:

> much of what is labeled as political geography is not very political. Often the political is read off from the economic or the cultural such that this or that economic interest or cultural identity, respectively, is more the subject of analysis than is the organizing of political agency in pursuit of this or that interest or identity. Under the influence of economistic varieties of neo-Marxism (particularly those of a heavily Leninist cast), ethnic identity politics, and essentialist versions of feminism the distinctively political (and the agency that comes with it) has disappeared into analyses that presume superorganic categories which determine political outcomes. (2003a: 604)

As these and earlier comments suggest, the gulf between feminist and traditional political geographers remains substantial. And the divide is likely to influence the shape of the subdiscipline for some time to come.

Organization of the Book

This book contains 28 concept chapters. Each chapter covers a key concept in political geography and is divided into three sections. In the first section, the concept is defined. In the second evolution in the concept's meaning and/or key debates are reviewed. Each chapter concludes with a case study showing how geographers have applied the concept in their research.

The 28 chapters are organized into six parts, each of which contains a group of related concepts. In Part I (Chapters 1–4), concepts of **statecraft** are outlined. Statecraft has been a central concern of political geography, and the concepts discussed here cover many of the subdiscipline's formative concepts, including governance, nation-state, democracy and sovereignty. In Part II (Chapters 5–8) concepts related to how political geographers understand **power** are discussed. The section includes chapters on hegemony, geopolitics, territoriality and superpower. Part III (Chapters 9–15) covers many of the formative concepts of the **modern era**. Since many of these concepts are temporally based – i.e. most cover a particular period within the modern era – they are ordered to reflect this. The section begins with chapters on colonialism/imperialism, political economy, ideology and socialism, and concludes with chapters on neoliberalism, globalization and migration.

Part IV (Chapters 16–18) is focused on the **interactivity** of political spaces. Special attention is given to connections and ruptures between political units. The section contains chapters on borders, scale, and regionalism. In Part V (Chapters 19–22), concepts related to the spatial manifestations of **violence** are considered. This part includes concept chapters for conflict, post-conflict, terrorism and anti-statism. The final part (Chapters 23–28) contains chapters, broadly linked under the heading of **identity**. Many of these concepts are related to the poststructural turn in the academy. They include nationalism, gender, citizenship, postcolonialism, the other and representation. It should be noted that the inclusion of representation in this section is indicative of the influence of poststructuralism in political geography. In traditional political geography representation was defined as the mechanism by which space was divided into political units for electoral representation. In the last twenty years, however, representation has come to encompass a wider set of concerns related to the ways in which identity groups are represented by the state (and other power brokers) in society and how such groups counter these representations. The decision to cover the concept of representation in this way was made with caution because the more traditional definition of representation remains an important concern in political geography. However, since traditional understandings of representation are covered in Part I, the chapter on representation was reserved for the emergent definition of the concept.

This is both a reference book and a source of in-depth knowledge on the concepts. The organization of the chapters into three discrete parts, for example, allows students to compare and contrast concepts as well as to go straight to select information about a concept. However, each chapter is also substantive enough to provide a foundation for students interested in learning about and using a given concept in their own research. It is the authors' hope that students will come away with an appreciation of the depth, complexity and relevance of political geography.

NOTE

1 Darwin developed the idea of natural selection to describe how certain traits became dominant in a species over time. Traits that allowed a species member to live longer, and thus reproduce more offspring, tended to become more common over time than 'weaker' traits. Lamarck, by contrast, argued that species variation was a product of 'will,' environment, and habit; substantial variations could take hold in one generation, unlike the more gradual change envisioned by Darwin. Lamarck's ideas were attractive to geographers because they enabled them to describe human differentiation as the result of human agency.

Part I
Statecraft

INTRODUCTION

Carolyn Gallaher

The first set of concepts considered in *Key Concepts in Political Geography* concern statecraft. The state is one of political geography's central units of analysis. Political geographers ask and answer a lot of questions about states. Some geographers focus on how states are formed and governed. Others analyse how state power is established, legitimized and resisted in the world system. Still others examine specific forms of state organization.

In this part of the book four concepts related to statecraft are considered. The first, the **nation-state**, is a central theme in geographic research. Geographers have long noted, for example, that nations – with the nation defined broadly as a group that sees itself linked by history, language and/or culture – do not always match the administrative boundaries of modern states. The nation-state is as much an ideal type as it is a realized entity in most places. Geographers have been at the forefront of examining the tensions that arise in places where national and state boundaries do not match.

The second concept discussed here is **sovereignty**. Geographers are interested in sovereignty because the concept encapsulates how a state gains and holds authority over the people living within its boundaries and the activities they engage in. The idea behind state sovereignty – that states are the only legitimate actors on the world stage – underpins the world system and the international laws designed to govern it. However, as geographers also note, globalization has undermined the durability of this view and some even suggest that the era of state sovereignty may end.

In the final two chapters here we examine statecraft more narrowly. Chapter 3 on **governance** examines the mechanisms by which a government accumulates capital and regulates the social polity. While states are meant to represent the interests of their citizens, at times citizens will contest the mechanisms by which they are governed. As such, geographers study not only how governance structures are organized, but also how they vary across social and geographic divides. In Chapter 4, a particular type of government structure – **democracy** – is considered. Although democracy is often held out as an ideal form of government, especially by western powers, geographers note that there is no universal definition of democracy. Governments can and do use a variety of mechanisms for ensuring some form of popular representation in government. And, each has its benefits and its disadvantages.

To get a feel for these concepts case studies are provided from across the globe, including South Africa, Mexico and the US.

1 NATION-STATE

Mary Gilmartin

Definition: A Concept's Two Parts

The term 'nation-state' is an amalgam of two linked though different concepts, nation and state. Nations are usually described as groups of people who believe themselves to be linked together in some way, based on a shared history, language, religion, other cultural practices or links to a particular place. States are usually defined as legal and political entities, with power over the people living inside their borders. In this way, states are associated with territorial sovereignty. The concept of a nation-state fuses together the nation – the community – and the state – the territory. In doing so, it provides us with a key unit of socio-spatial organization in the contemporary world. In defining the nation-state, it is important to consider its two separate components as well as the relationship between nation and state. A nation, according to Anthony Smith, is a named human population 'sharing an historic territory, common myths and historical memories, a mass, public culture, a common economy and common legal rights and duties for all members' (in Jones et al. 2004b: 83). Smith's definition points to a number of commonalities, around territory, culture, history and memory, which may suggest that there is an essential quality to a nation. An essentialist understanding of a nation (sometimes called primordialism) suggests that it has always existed, and that it has an unchanging core.

An essentialist view of the nation is strongly contested by those who see nations as socially constructed. For example, Benedict Anderson famously argued that nations are imagined communities, because 'the members of even the smallest nation will never know their fellow members, meet them or even hear of them, yet in the minds of each they carry the image of their communion' (Anderson 1983: 15–16). Nations, in this way of thinking, come into being to serve particular purposes, often economic or political (Storey 2001: 55). This approach to nation formation may either be perennialist or modernist (the modern era is usually defined as beginning with the industrial revolution). A perennialist theory of nation formation suggests that the nation is rooted in pre-modern ethnic communities. In contrast, a modernist theory of nation formation suggests that processes associated with the modernist period – such as the development of states, the advent of mass literacy and education, or the spread of capitalism – led to the creation of nations.

The second component of the concept of the nation-state is the state. John Rennie Short observed that one of the most important developments of the twentieth century was 'the growth of the state' (Short 1993: 71). Short commented both on the increase in the number of states, from about 70 in 1930 to over 190 in 2007, and on the growth of state power. The increase in this period in the number of states is closely linked to decolonization. As empires were dissolved, particularly after World War II, imperial spatial organization (where territories were governed from the centre of the Empire, for example London) was replaced, to a large extent, with a state-based system of spatial organization. Many of these new states were based on European models, with a strong emphasis on territoriality and on the management of people and resources. Contemporary states have power and influence over both internal and external relations. Internally, the state works to gather revenue, maintain law and order and to support the ideology of the state. Externally, the state works to defend its borders and territory and to maintain favourable political and economic relations (Jones et al. 2004; Short 1993: 71) (see Figure 1.1).

The nation-state is an ideal type: it suggests that the borders of the nation and the borders of the state coincide, so that every member of a nation is also a member of the same state, and every member of a state belongs to the same nation. In practice, this is impossible to achieve. The result is a variety of combinations of nation and state. One combination is states which contain many nations, such as Spain, with minority nations such as Basque and Catalan (see Figure 1.2). Another is nations spread across more than one state, such as the Irish nation in the Republic of Ireland and also Northern Ireland (part of the United Kingdom). A third combination is that of nations without states: the Basque nation may be defined in this way, as may the Kurdish nation, living in Turkey, Syria, Iran and Iraq.

There are clear disagreements over how a nation-state comes into being. However, these disagreements do not extend to the influence of

Figure 1.1 A fortified portion of the border between Mexico and the US

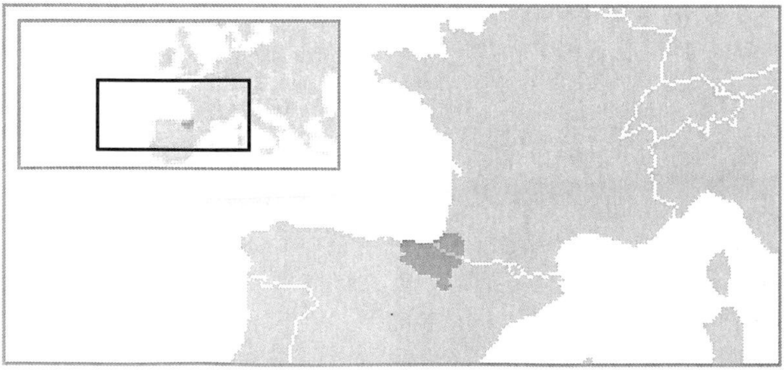

Figure 1.2 Map of Basque provinces in Spain and France

the concept and its ability to galvanize people into action. This happens through nationalism, described by Anthony Smith as 'an ideological movement for attaining and maintaining autonomy, unity and identity on behalf of a population deemed by some of its members to constitute an actual or potential nation' (in Storey 2001: 66; for more detail see Chapter 23). Smith suggests that there is a distinction between ethnic and civic nationalism, where ethnic nationalism focuses on shared ethnic identification and commonalities, while civic nationalism focuses on shared institutions. The distinction between ethnic and civic nationalism suggested by Smith creates a hierarchy of nationalisms. This has implications for how nations and states are understood, with the creation of categories of 'failed' (and, by association, successful) states in the contemporary world.

Despite their obvious differences, the terms nation, state and nation-state are often used interchangeably. It is important to acknowledge this slippage. For example, much work in political geography highlighted the state, but was based on an implicit assumption that the state was also a nation – in other words, that its population shared a particular national identity. As such, the term 'state' implied national cohesion, but often served to mask conflict at subnational levels: between ethnic or racial groups, between regions, or around issues of power or ideology. In a similar vein, the use of the term state implied a form of civic nationalism, which again served to reinforce global hierarchies, even though the territory may well have been in the process of ethnic nation-building. The politics of naming is significant, and the assumptions underpinning the categorizations of nation, state and nation-state should always be interrogated.

Evolution and Debate: Is the Nation-state Relevant Any More?

The nation-state is one of the building blocks of political geography. Early political geographers, such as Friedrich Ratzel and Halford Mackinder, paid particular attention to the nation-state: Ratzel in his conceptualization of the state as a living organism that needed to grow in order to survive; Mackinder through his articulation of the state as a place where social and political goals could be pursued (Agnew 2002: 63–70). The state remained at the centre of political geography, so much so that Peter Taylor has argued that the focus on the state as a spatial entity distinct from social conflict led to an innate conservatism in the discipline (2003: 47). In other words, political geographers were so concerned with privileging the state as an entity that they failed to adequately investigate the state as a site of contestation, for example between different ethnic groups living in the state.

This lack of attention to contestation within the nation-state has been addressed in recent years, particularly through a greater concern with questions of identity. On one level, this has been addressed through a focus on the process of nation-building, with particular attention to monumental, memorial and other symbolic landscapes (see Johnston 1995 and Whelan 2003 for a discussion of this process in Ireland). In recent years, political geographers have also been more attentive to questions of gender, race, ethnicity, class and sexuality, highlighting ongoing contestations over the definition of the nation-state, as well as challenges to the processes of exclusion that underpin national identities. For example, feminist political geographers have highlighted the gendered nature of nation-states and national identity and have argued for a deeper engagement with the ways in which feminized and apparently private spaces, such as the household, are central to how nation-states are imagined and work (see Staeheli et al. 2004). Similarly, recent work on sexuality within political geography has highlighted the ways in which nation-states are often heteronormative, with national identities constructed around an assumed heterosexuality. This attention to identity has most recently been articulated in relation to citizenship (for example, see the discussion of sexual citizenship in *Political Geography* 25: 8, which attempts to move the concept of citizenship beyond the political, and argues that sexuality is part of citizenship). The concept of citizenship, particularly in relation to individuated rights and responsibilities, has been the focus of much recent research on states within political geography. (See Chapter 24 for a more detailed discussion.) Postcolonial theory has been used by geographers to question the exclusionary practices of nation-states after colonialism. (See Chapter 25 for a more detailed discussion.) In addition to highlighting debates about processes of inclusion and exclusion and resulting conflicts within the boundaries of the nation-state, political geographers have also started to engage more broadly with questions of governance within the nation-state. This has included a focus on local scales, such as the changing forms and functions of local states in a globalizing and neoliberal world (see Jones et al., 2004b Chapter 4, for an overview). This has also included a focus on protest and resistance movements, as well as the state's responses to such movements (see Herbert 2007).

More fundamentally, however, the nature and existence of the nation-state has itself come under scrutiny. For some commentators, the nation-state is an anachronism, superseded by supranational organizations such as the United Nations and the European Union, and by processes such as globalization. John Agnew has suggested that 'the modern territorial state is now is question in ways that would have been unthinkable even twenty years ago' (2002: 112). Agnew highlights globalization, global migration, the collapse of the 'strong states' of the Soviet Union, the growth of supraregional and global forms of governance and the increase in ethnic and regional conflicts within states to support his assertion. The relationship between the nation-state and globalization has received particular attention. One school of thought is encapsulated in Kenichi Ohmae's comment that 'traditional nation-states have become unnatural, even impossible business units in a global economy' (in Jones et al. 2004b: 51). In contrast, others suggest that the nation-state remains important despite globalization (see Yeung 1998). Similar ambivalence is evident in discussions of global migration, with some arguing that the so-called 'age of migration' has led to significant numbers of transnational migrants, who maintain strong networks and links with their countries of origin as well as their places of residence. Their presence and their activities, it is suggested, challenge state and national borders and ideologies (see Nagel 2001 for an overview). Other commentators suggest, however, that the scale of global migration has led to a tightening up of state immigration policy and an intensification of border controls and surveillance. This ambivalence is also present in discussions of global terrorism and global social movements, and in debates over the extent to which states can or cannot, as the case may be, contain and control terrorist or protest activities within their borders. This has suggested the concept of a failed state, described by some commentators as a state that is incapable of asserting authority within its own borders, but seen by other commentators as a neocolonial concept applied primarily to former colonies. In short, the nation-state in the contemporary world is, despite its ubiquity, a contested concept, and political geographers are central to debates over its contested meanings.

Case Study: South Africa after Apartheid

South Africa provides an interesting site for the study of the nation-state. During the apartheid era in South Africa, the population of the country was divided on racial and ethnic lines into separate territories. Under apartheid, South Africa was clearly not a nation-state,

Figure 1.3 Voortrekker Monument, Pretoria, South Africa

but consisted of a number of nations – racially and ethnically defined – within one state. Those nations were socially constructed. As an example, Crampton has written of the importance of the Voortrekker Monument in Pretoria in articulating a nationalist Afrikaner identity in the 1940s (Crampton 2001). The Voortrekker Monument was intended to celebrate the Great Trek of Afrikaners into the interior of South Africa in the 1840s, and Crampton considers the inauguration of the monument in 1949 as part of a broader Afrikaner nationalist project (see Figure 1.3). As apartheid ended, a variety of groups argued that South Africa needed to construct a new identity for the state through a process of nation-building. President Nelson Mandela, for example, called for a 'rainbow nation', and described his vision for this new South Africa as follows:

> In centuries of struggle against racial domination, South Africans of *all colours* and backgrounds proclaimed freedom and justice as their unquenchable aspiration. They pledged loyalty to a country which belongs to all who live in it ... Out of such experience was born a vision of a free South Africa, of a nation united in diversity and working together to build *a better life for all*. (in Ramutsindela 2001: 74, emphasis in original)

However, the economic realities of post-apartheid South Africa posed obstacles to achieving this vision. In particular, the higher levels of poverty and unemployment and the lower levels of education among black communities meant that building a better life for all effectively meant improving conditions for black South Africans. As a result, politicians in South Africa, like President Thabo Mbeki, became more concerned with the black or African nation than the rainbow nation (Ramutsindela 2001). Many white South Africans believe that such actions discriminate against the white community and, as a result, question whether there is a place for them in the new South Africa. White South Africans – in particular, Afrikaners – have reacted in a variety of ways. Some have cooperated with government policies, for example by participating in the country's land reform programme. Others resist government policies, and have begun to draw again on an Afrikaner national identity, linked to land, history and a shared struggle (Fraser 2008). These various nation-building projects have been compromised by the adherence of the South African post-apartheid nation-state to free market principles and neoliberalism (Peet 2002). In these ways, post-apartheid South Africa highlights the complexity of the concept of nation-state as well as its contested nature.

KEY POINTS

- The nation-state is the key unit of socio-spatial organization in the contemporary world. The term links together the nation – a community of people with an assumed connection – and the state – a legal, political and territorial entity.
- Within political geography, the nation-state has been a key area of focus. However, political geographers were historically more concerned with the state, and paid less attention to conflicts within states.
- In recent years, the study of the nation-state within political geography has expanded to address questions of identity, citizenship and governance, as well as the place of the nation-state in a globalized world.

FURTHER READING

Agnew, J. (2002) *Making Political Geography*. London: Arnold.
Anderson, B. (1983) *Imagined Communities: Reflections on the Origins and Spread of Nationalism*. London: Verso.

Storey, D. (2001) *Territory: The Claiming of Space*. Harlow: Pearson Education.
Taylor, P. (2003) 'Radical political geographies,' in J. Agnew, K. Mitchell and G.Toal (eds), *A Companion to Political Geography*. MA and Oxford: Blackwell. pp. 47-58.
Whelan, Y. (2003) *Reinventing Modern Dublin: Streetscape, Iconography and the Politics of Identity*. Dublin: University College Dublin Press.

2 SOVEREIGNTY

Carl T. Dahlman

Definition: A Multi-Faceted Concept

The ultimate authority to rule within a polity is known as sovereignty. Historically, ultimate authority within a polity was located in the person of the sovereign, a monarch whose rule was vested by divine right or local custom, and often by a good deal of force. In feudal Europe, a monarch's rule was not conceived territorially, except what he owned personally. Rather, his authority extended from the fealty of those loyal to him and they, in turn, possessed the land; when allegiances shifted, so too did the area under sovereign rule (Figure 2.1).[1]

The modern territorial state system began to take shape under the Capetian dynasty (10th–14th centuries) and was largely consolidated by the time of the Westphalian peace in 1648. This system defines sovereignty as the right to rule a territorially bound state based on two types of authority (Sassen 2006). The first pertains to internal or domestic sovereignty, the authority to rule within a delimited territorial state, which requires that the monarch's subjects recognize his right to rule (Wendt 1999: 206–211). Complementary to internal sovereignty, a second external or international legal sovereignty is the right of a sovereign government to rule its territory without external interference. This is predicated on the recognition of sovereignty by other sovereign entities, which since 1945 has been largely routinized in admission into the United Nations.

Contemporary norms of sovereignty remain bound to these two forms of recognized authority, and polities are thought to possess sovereignty absolutely or not at all (Williams 1996: 112). In effect, sovereignty comprises the legal personality of the modern state, an amalgam of government, territory and people, which thus possesses rights, powers, jurisdiction, etc. In this rendering, sovereignty may be violated or derogated but it is not divisible or additive – a state cannot gain some sovereignty at the expense of another, as is possible with territory (Brownlie 1998: 105–106). Stripped bare, legal sovereignty should be

Figure 2.1 Although the monarchy had long been limited by the time Queen Victoria took the throne, she remained a powerful example of a remnant system of personal sovereignty in Europe. She ruled over the major expansion of the British Empire and was officially both Queen of the United Kingdom and Empress of India. (Engraving from 1897, courtesy of the National Archives and Records Administration).

considered separate from the issues of power, control and jurisdiction that most people imply when they invoke the term. This is analytically useful since sovereigns have actually possessed differing abilities within their polities throughout history, from besieged figurehead to totalitarian despot. The empirical variability of what sovereignty means in practice reminds us that it is more an ideal or legal abstraction than what exists in reality and most authors thus imply a wide range of meanings when they invoke the term.

Stephen Krasner usefully identifies four meanings of sovereignty that are differentiated by varying degrees of authority and control and which also highlight the implicit geographical construction of the international state system (Krasner 1999). First, domestic sovereignty includes both authority and control but only within a state. While internal political authority (the right to rule) may be constituted in various ways, it is distinct from whether a sovereign entity exercises effective control throughout state territory. Second, international legal sovereignty dwells on the issue of the authority of the state, namely its government, within the international realm where sovereign states interact. This authority ultimately rests on the recognition of state sovereignty by other sovereign states and, importantly, this legal personality can survive territorial and internal changes. Third, Westphalian sovereignty implies the territorial organization of the state as an inviolate realm, free from intervention by other states. The norm of non-intervention is often at the crux of what is implied by 'violations of sovereignty', and although states may violate their own Westphalian sovereignty by invitation (asking another state to come to their defence), this does not diminish their international legal sovereignty. Fourth, interdependent sovereignty relates wholly to the state's control of its borders and its exposure to external influences. Although Krasner focuses on commercial transborder flows of capital, labour and goods, this could extend to other transnational movements, such as ideological and cultural diffusions. It is worth noting that these different forms of sovereignty may have transformative effects on one another but they need not vary in direct proportion to each other.

Evolution and Debate: Locating Sovereignty

As the above discussion implies, sovereignty is both an ambitious and a vague term. On the one hand, it is thought to provide the fundamental framework for the political and geographic organization of modern international politics. On the other hand, its meaning is both highly contextual and often not fully stated, rendering the concept less analytically reliable than we would expect.[2] The concept also presents numerous conundrums that authors have examined in some detail. For example, if internal sovereignty is dependent on the unlimited right of a ruler, then what happens when states are governed as representative democracies? The fragmentation of sovereignty within a modern governmental apparatus on top of democratic norms regarding popular sovereignty mean that a Hobbesian understanding of sovereignty (in which sovereignty is possessed by someone with special rights and separate from the rest of the polity) is simply not useful to understanding today's political world. Instead, modern democratic government suggests that sovereignty is something that can be unbundled into different competencies and even shared among various institutions within a state, e.g. the people, local political units or federal structures. Furthermore, if the right of non-interference means that the internal affairs of sovereign states are inviolate, then how do we square this with international norms protecting human rights, which might require an intervention (invasion) to protect victims and/or arrest violators? And what about situations where the actual area of sovereign control exercised by a state is smaller than its legal territory, and in the gap are sheltering threats to it or a neighbouring state?

Normative considerations aside, students of actually existing politics must engage with the fact that power, territory and rights are increasingly reorganized into spaces other than states, challenging our understanding of a world based on sovereign territorial actors (Luke 1996: 494). To this end, it is useful to consider a series of geographical arguments that fundamentally reshape how we understand sovereignty. First, conventional approaches to sovereignty tend to view territory as merely the stage, container or resource of sovereign power rather than fundamental to its operation. It is necessary to grasp the geographic concept of territoriality as a historical explanation of how sovereignty has been organized through social practices that use bounded spaces as a medium of power and identity. Sovereignty is therefore an evolving spatial practice as states try to respond to changes and challenges in domestic and international economic, social and political life.

Second, seen through the lens of territoriality, it becomes easier to identify a further distinction between *de jure* (legal or recognized) and *de facto* (actual) sovereignty. The difference is significant since recognition of sovereignty is not the same as effective control of that territory. Governments in exile are an important example of why this matters. Likewise, international legal sovereignty might not be matched by a commensurate recognition of domestic sovereignty, as in the case of many decolonized states during the twentieth century (Murphy 1996).

Third, John Agnew takes this distinction a step further to argue that '*de facto* sovereignty is all there is', identifying four ideal-type regimes of effective sovereignty according to the strength of state authority and its territorial reach. Strong state authority within a bounded state territory represents the 'classic' type of sovereignty, while a weak central authority beholden to outside powers and challenged by separatist forces represents the opposite 'imperialist' type. Agnew identifies the European Union as an example of 'integrative' sovereignty, in which state authority is weaker by having vested certain competencies in a regional institution. The fourth type, 'globalist', is exemplified by the United States in having retained a strong central authority while extending its influence beyond its borders by getting other states to sign on to economic and political agendas for which the US serves as hegemon (Agnew 2005).

Agnew's argument recognizes a fourth point, voiced by others: that states are never as much in control of their affairs and territory as the ideal of sovereignty implies (Krasner 1999; Williams 1996). But whereas Agnew's formulation of effective sovereignty concentrates on states, there are a variety of other forces in world politics that are not sovereign actors yet challenge state sovereignty. Most commonly, these forces are manifestations of globalization, where multilateral treaty organizations, international trade, labour migrations and transnational cultural diffusions challenge the basis of state sovereignty. Neoliberal economic policy, for example, is thought by some to present a turning inside-out of state authority as once-sovereign powers submit to decisions made by bodies such as the World Trade Organization or the imperatives of multinational corporations. International action in defence of human rights suggests perhaps a more fundamental transformation of sovereignty since it not only violates the non-intervention norm but seeks to reform what counts as the legitimate practice of domestic sovereignty.

This leads to a final point – many territories are effectively ruled not by states but by other non-state entities, such as guerilla/paramilitary groups, criminal gangs and illicit trade networks. In most cases, non-state actors merely limit the *de facto* sovereignty of a state over parts of its territory. That is, they limit a state's ability to exercise effective control over its entire territory, even though the uncontrolled territory is often still recognized, both domestically and internationally, as part of the state. In some cases, however, non-state actors may achieve not only effective territorial control but also effective ruling authority, which local populations may recognize as a *de facto* domestic sovereignty. The challenges to state sovereignty introduced by territorialized non-state actors may similarly apply to occupying powers and international administrations established to temporarily govern a failed state, as in Kosovo or Iraq. But this raises doubts about the legitimacy of domestic sovereignty once a domestic government is restored. In Iraq, for example, the United States remains an occupying power although formal sovereignty has been handed over to a recognized Iraqi government, raising the question of whether the US would indeed withdraw if asked by the Iraqi government. Like globalization, these forces challenge sovereignty and cause us to re-examine what we mean when we think of states. At the same time, sovereignty has proven to be an important normative concept capable of historical transformation that may yet survive along with the state, albeit in response to changing expectations of what ultimate authority within a state actually entails (Taylor 1994).

Case Study: Bombing Serbia to Save Kosovo

Humanitarian intervention, namely armed intervention to defend the human rights of another country's population without that government's permission, raises hotly debated questions about sovereignty. Although such actions may appear similar to wars of conquest, they are predicated on a sense of moral responsibility to stop continued abuses rather than to gain territory from the target state. Moral and ethical reasons for violating another state's sovereignty, however real and compelling, have not yet been established in international law so these interventions generally occur in a legal grey area. In fact, states that undertake humanitarian interventions are careful in their justifications to avoid the claim of customary right to such actions and instead cite more narrowly construed national security interests (Holzgrefe 2003: 47–49). Leftist critics of humanitarian intervention often accuse western powers, especially the United States, of using such opportunities in the pursuit of empire, hegemony, regime change, or merely to exercise their might for show. Interestingly, they share a perspective commonly voiced by states with poor human rights records, e.g. Serbia. These criticisms in any case typically ignore the substance of the human rights issue and are not discussed here.

The Kosovo crisis from 1998 to 1999 provides an excellent case study through which to examine current issues affecting state sovereignty (Figure 2.2). The case of Kosovo is situated within the larger context of

 Figure 2.2 Map of the Socialist Federal Republic of Yugoslavia (based on United Nations map 3689, June 2007)

the break-up of the former Yugoslavia and extensive human rights abuses instigated by the leaders of that state as it was falling apart. Although the larger dissolution of Yugoslavia presents interesting questions about state sovereignty and territoriality, the focus here is on Kosovo, which was a distinct region within the Republic of Serbia, which was a part of the larger Yugoslavia. In particular, it raises three questions regarding Serbian sovereignty over Kosovo. The first relates to the basis of that claim and the subsequent independence movement that led to the crisis. The second question concerns the challenge to sovereignty raised by humanitarian intervention, namely the effect of the NATO bombing campaign against Milošević's Yugoslavia, the country left over after the independence of Slovenia, Croatia, Bosnia and Herzegovina, and Macedonia. The third question concerns the international recognition of Kosovo as an independent state.

Serbian sovereignty in Kosovo has been repeatedly challenged, beginning, in fact, with pre-Westphalian claims to the area. Kosovo had been part of the early Serbian empire before the Ottomans conquered it in the fourteenth century, after which it slowly became more ethnically Albanian Muslim. The modern Serbian Kingdom gained the province by the Treaty of London (1913) and formally annexed the territory with the Allied defeat of the Ottomans in World War I. Annexed to Albania by Fascist Italy during World War II, it was returned to post-war Yugoslavia, whose federal socialist government did little to help the province. Agitation for ethnic Albanian rights was finally addressed in 1974 by a new Yugoslav Constitution that granted Kosovo the status of an autonomous province. Still, Kosovar Albanians were not recognized among the constituent nations of Yugoslavia, such as Serbs, because Albanians were considered to have a separate national homeland outside Yugoslavia. Instead, they were viewed as minorities within Yugoslavia with fewer rights than the constituent peoples in whose name Yugoslavia claimed sovereignty. As an autonomous province, however, Kosovo had a local government, judiciary, police and school system, as well as a sizeable Serb minority (18.3 per cent in 1971, falling to below 10 per cent in the 1990s) (Statistical Office of Kosovo 2008: 7).

In his rise to power, Yugoslav President Slobodan Milošević sought to exploit inter-ethnic tensions in Kosovo, which fuelled support among Serbs for the reactionary Serb nationalism that helped him attain office. Playing up a political discourse of Serb victimhood at the hands of Muslims, Milošević made symbolic use of Kosovo's historical setting as the location of a major battle between the early Serb empire and Ottoman Turks, a battle that had its 600th anniversary in 1989. In the course of 1989, Milošević took control of local Kosovo government, revoked its autonomous status, and began a process of discriminatory Serbianization in the province (Independent International Commission on Kosovo 2000: 33–42). As Milošević was gaining unchecked powers, the republics of Slovenia, Croatia, and Bosnia and Herzegovina sought independence from Yugoslavia, leading to a series of wars that lasted from 1991 to 1995. Kosovo, meanwhile, remained under Milošević's rule. During the 1990s, local Kosovar Albanians organized a parallel government that grew into a separatist movement, adding to growing unrest in the province. Despite the attention of international diplomats working to stop the wars in the other republics, the treaties of the 1990s did not address the issue of Kosovo. Out of this period came a guerilla organization, the Kosovo Liberation Army (KLA), that in 1996 began attacks it hoped would provoke international interest and ultimately lead to Kosovo's independence, directly challenging Serbia's sovereign claims to the province (Judah 2008: 77–80).

From late 1997 until October 1998, the KLA made confrontational shows of force while declaring its strongholds in the province 'liberated'. The Yugoslav police, Serbian special police, and Serb paramilitary units – similar to the Serb formations that fought in Croatia and Bosnia – began a series of attacks on KLA positions. The Serbian forces also attacked ethnic Albanian civilians and began depopulating areas under its control. In response, the KLA attacked Serb villagers in some areas. The violence caused many to flee and by the end of 1998 more than 250,000 persons, mostly Albanian but including many Serbs, had left their homes. The already large displacements created during the first phase of the conflict were dwarfed after October when the Serb forces under Milošević's direction took direct measures to ethnically cleanse Kosovo of ethnic Albanians (NRC 2004). The growing crisis, especially in the wake of revelations regarding the scope of ethnic cleansing and genocide during the Bosnian war, brought intense international scrutiny and concern for the human rights of civilians in Kosovo.

The international response to the crisis in Kosovo raises a second issue, that of humanitarian intervention contra Yugoslav and, later, Serbian sovereignty. The United Nations Security Council in March 1998 adopted Resolution 1160 calling on the parties to seek a political solution for ending the violence and for giving the province 'an enhanced status' within the Yugoslav state, including autonomy and meaningful self-administration. Although Resolution 1160 affirmed the 'sovereignty and territorial integrity' of Yugoslavia it also demanded that Yugoslavia withdraw some of its forces and accept an Organization for Security and Cooperation in Europe (OSCE) observer mission while placing the country under an arms embargo. It explicitly put the events in Kosovo under the jurisdiction of the International Criminal Tribunal for the Former Yugoslavia (ICTY), which was investigating war crimes in Croatia and Bosnia. The conflict grew and attacks on civilians increased during the months that followed, and in September the Security Council passed Resolution 1199 that called for a ceasefire, continued negotiations, open access for diplomatic and authorized international agencies, and the withdrawal of Yugoslav forces used for civilian repression. In October, the NATO council authorized the use of force against Yugoslavia if it failed to comply with Resolution 1199, causing Milošević to accept the Security Council's terms. Yugoslavia's agreement to end the conflict in Kosovo was affirmed by UN Security Council Resolution 1203, and Resolution 1207 required Yugoslavia to cooperate with the ICTY, which was ready to pursue war crimes investigations and prosecutions for the events in Kosovo.

Although government and KLA forces appeared at first to respect the ceasefire, renewed KLA attacks led to Yugoslav army action soon after the October agreement. From October 1998 to June 1999, government forces under Milošević's direction armed local Serb militias to ethnically cleanse Albanian villages, summarily executing civilians in the process. Unlike in Bosnia, where verification of widespread human rights abuses was at first difficult to obtain, US and European diplomats and OSCE monitors were in Kosovo and reported immediately on Serbia's excessive use of force and human rights violations against Kosovo's civilians (Independent International Commission on Kosovo, 2000: 67–83). An attack on civilians in the village of Recak was immediately verified and condemned by the OSCE observer mission, helping to accelerate the international response. When the chief prosecutor of the ICTY attempted to enter Kosovo, however, she was prevented from doing so by the Serbian government; the head of the OSCE mission was then declared *persona non grata* – an anodyne demonstration of sovereignty (Independent International Commission on Kosovo 2000: 75–83).

February negotiations in Rambouillet in France failed to produce a settlement between the Milošević regime and the KLA. The OSCE evacuated its mission in March, after which the conflict grew worse, followed by NATO airstrikes against military targets in Kosovo and strategic sites elsewhere in Yugoslavia. The air campaign lasted from 24 March to 10 June and caused enormous damage to Yugoslav military sites and civilian infrastructure, in addition to killing hundreds of civilians. The actions by NATO were taken without explicit UN Security Council authorization; the alliance instead cited the implicit authority to apply force according to Security Council resolutions which were adopted under the UN Charter's Chapter VII powers. The government's deliberate ethnic cleansing of Albanians continued during the airstrikes and was visible in the exodus of displaced persons from Kosovo seen by global TV viewers. By June, the government campaign had displaced about 1.45 million people, about 90 per cent of all Kosovar Albanians (NRC 2004). Milošević attempted to maintain his country's sovereignty over Kosovo by forcibly removing its 'non-sovereign' majority population of ethnic Albanians.

A third question of Yugoslav and, since 2003, Serbian sovereignty over Kosovo was raised immediately following the negotiations that stopped the NATO strikes.[3] The Security Council adopted Resolution

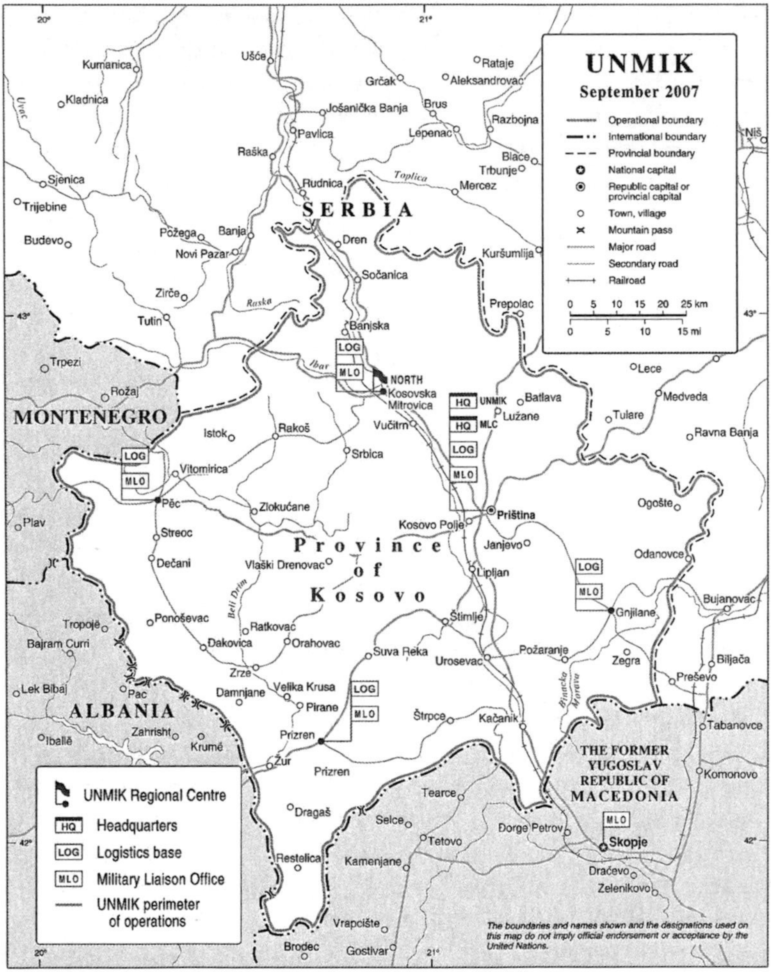

Figure 2.3 Map of United Nations Mission in Kosovo (UNMIK), key sites. Note the clear distinction between the province boundary, international borders, and the UNMIK perimeter of operations (based on United Nations Map 4133, June 2007)

1244 that established a *de facto* protectorate in Kosovo, effectively separating it from the rest of Serbia (Figure 2.3). The structure of the international administration in Kosovo has two parallel elements: a military peacekeeping operation (KFOR, or Kosovo Force) and a civilian operation under UN auspices including OSCE and EU involvement.

Resolution 1244 also set forth the expectation that international negotiations would resolve the final status of Kosovo. These negotiations did not begin in earnest until 2006, after which the UN special envoy outlined a proposal for what implicitly appeared to be independence for Kosovo. Serbia's government remained steadfastly opposed to such an outcome, proposing instead a final status 'more than autonomy but less than independence'. Serbia has been supported on this matter by Russia, which has blocked any further Security Council Resolution that would replace 1244.

Towards the end of 2007, it was clear that negotiators would not reach an agreement on Kosovo's final status that would be acceptable to both Serbian and Kosovar governments. What remained of the negotiations was UN Envoy Martti Ahtisaari's draft status proposal that had been blocked by Russia at the UN Security Council. The plan nonetheless shaped the international expectation of Kosovo's eventual independence that began to move forward without a new resolution by the UN Security Council. The Prime Minister of Kosovo, a former KLA military commander, indicted that Kosovo would make a 'coordinated declaration' of independence on advice from the United States and the European Union. This took place on 17 February 2008, and the United States, France, the United Kingdom, Germany and several other states quickly recognized Kosovo's sovereign independence. Within the year Kosovo had been recognized by 22 EU member states and about 30 other states – still a minority of the 192 members of the United Nations. The European Union itself has no formal means of recognizing Kosovo's independence, but it has supported independence by offering to replace the UN international administration with relevant EU support missions. Although Russian objections to Kosovo's independence might relate to Moscow's own interests in the Caucasus, the largest issue by far facing Kosovo is Serbia's continued claim to the province. Serb enclaves that formed after the 1999 war and riots in 2004 are now *de facto* pockets of Serbian rule on the territory of Kosovo. Belgrade has sponsored local parallel Serb administrations in these enclaves that continue to provide identity cards, property documents and other basic services to its residents. These enclaves also hosted polling locations during the Serbian parliamentary election held shortly after Kosovo declared independence. Thus even as Kosovo achieves wider international recognition, its sovereignty is challenged by obstacles to membership of the UN and Serbia's continued claim to at least part, if not all, of the province.

KEY POINTS

- Sovereignty is the ultimate authority to rule. In practice this authority has changed over time in response to attempts to limit it, e.g. democratization, economic liberalism and humanitarianism.
- Contemporary sovereignty is closely entwined with a state's territory. Although they differ in important ways, state territory is the fundamental medium of sovereignty.
- Interpretations vary as to what state sovereignty implies for contemporary political rule, especially with regard to human rights.

NOTES

1 The gendered language here reflects the widespread use of agnatic primogeniture in the succession of title and land, though female inheritance was not everywhere barred during the Middle Ages.
2 For an example of how the term is used in very different ways in the US academic context alone, see the provocative article by Sinclair and Byers (2007).
3 In 2003, the Federal Republic of Yugoslavia became the State Union of Serbia and Montenegro, followed in 2006 by the dissolution of the union, leaving Serbia and Montenegro as separate sovereign countries.

FURTHER READING

Agnew, J. A. (2005) 'Sovereignty regimes: territoriality and state authority in contemporary world politics,' *Annals of the Association of American Geographers*, 95(2): 437–461.

Judah, I. (2008) *Kosovo: What Everyone Needs to Know*. New York: Oxford University Press.

Krasner, S. D. (1999) *Sovereignty: Organized Hypocrisy*. Princeton, NJ: Princeton University Press.

Mbembe, A. (2000) 'At the edge of the world: boundaries, territoriality, and sovereignty in Africa,' *Public Culture*, 12(1): 259–284.

Sassen, S. (2006) *Territory, Authority, Rights: From Medieval to Global Assemblages*. Princeton, NJ: Princeton University Press.

3 GOVERNANCE

Peter Shirlow

Definition: Managing State Affairs

The term 'governance' refers to the mechanisms by which a government establishes a regime of accumulation and social regulation.[1] There is a substantial literature that examines how the concept of governance might be defined. The starting point with regard to understanding governance is usually the effect of government or supranational institutions upon the delivery of policy, practice and accountability. Governance should be based upon interlinked relationships that reflect upon tradition, uphold state and civic institutions and determine, via democratic accountability, how power is both exercised by elected representatives and statutory agencies in order to respond to the rights and obligations of citizens. (Bennett 1998). Most of the discussion of governance usually emphasize the following broad definitions:

- the processes by which governments are selected, scrutinized and modified;
- the capacity of government to generate, establish and apply policy and legal procedures;
- the systems and methods of interaction between the executive, the legislature and the judiciary;
- the relationship between the state, economy and civil society and the regulation of economic regimes and of social authority and practice;
- the instruments and apparatus by means of which citizens and vested interest groups identify and characterize their welfare and interests to institutions of authority, and vice versa.

In considering these broad issues more carefully two variables are especially important. First, governance requires the exercise of power. Various types of power are used in the application of governance, including establishing political and legal authority, creating ideological representations, maintaining regulation over domestic markets and controlling the means of violence. Issues of accountability and resistance to such power are also considered.

All governments or supranational institutions must use a combination of these types of power to govern (Brenner 1998). In non-representative states, governance is inadequate and representative of the will of 'the few', whilst in more democratic societies the capacity to govern is based on public accountability and the consent of citizens. Second, and crucially within geographic deliberation, is the idea of scale and how to distinguish between issues of governance at the macro and micro levels. (For more information on scale see Chapter 16.) At the macro level geographers are concerned with:

- creating an explanation of evolving international and global power structures and the shifting nature and structure of authority frameworks;
- considering how accepted democratic norms such as accountability, transparency, democratic representation and participation are (or are not) embedded in emergent and transforming global policy-making institutions;
- outlining the shift from national government to multilayered global governance and the frameworks created within private, public and assorted governance arrangements that stretch beyond the state;
- promoting fair and equitable modes of global governance.

Micro issues of governance include:

- The devolution of power to regions and other subnational units (assemblies, councils);
- Government policies aimed at sub-national arenas (local social inclusion networks and bodies, government-sponsored social economy projects, non-government organizations);
- Policies tailored specifically to ethnic or marginalized groups;
- Informal institutions of governance (churches, community sector) and other non-state institutions (private companies, voluntary organizations).

The emergence of governance as a definitional research term in the 1970s and 1980s was centred upon determining better ways to comprehend the complexities of government and democratic accountability (Harding 2000). Since then the emphasis has been more upon moving beyond an understanding of governance that merely comprises an examination of political traditions and the evolution of institutions. This shift has favoured a more accurate understanding of the relationship between governance and the exercise of power and the processes through which citizens are involved in or excluded from evolving modes of governance. Governance thus increasingly deals with issues relating to the mechanisms required to negotiate between various and competing interests. In addition, the term is also linked to understanding the manner in which governance, via the institutions of power and interest, can maintain and deepen empowerment and ensure that citizens own the process of governance. Evidently the right of the 'people' to participate in and make decisions with regard to the affairs of their nation, region and community is seen as critical in democratizing both the state and wider society (Monbiot 2000).

One of the key points of view regarding the success of new modes of governance is that they should be characterized by heterarchy (inclusion) rather than by hierarchy (domination). This should be achieved by developing horizontal modes of governance across space via an amalgam of actors within both the public and private spheres of human activity. The key to developing heterarchical governance is the involvement of 'relevant' stakeholders (Jessop, 1990). Smismans (2006) argues that such notions of governance are linked to a normative democratic claim that praises the particular participatory features of 'new governance' as compared to 'old governance'. Furthermore, he states that: 'More horizontal and heterarchical governance does not mean automatically more participatory governance in normative democratic terms' (2006: 1).

Governance is more than the act of government, given that it stretches beyond institutional frameworks and into areas including institutions and actors such as quangos, pressure groups, lobby groups, social movements and non-governmental organizations. The recent broadening of the concept of governance reflects the reorganization of the state and a more reasoned analysis that the state is not omnipresent with regard to the reproduction of society and societal norms. As non-state organizations have proliferated, such as training agencies and local development authorities, the overall structure of governance has altered. This alteration in the structure of democracy is seen by those who promote it as localizing democracy, whilst others view the rise of such bodies as undermining accountability concerning state actions (Peck 2002). Therefore part of the study of governance has been dedicated to the nature of this transition and the 'hollowing out of the state'.

This 'hollowing out of the state' has engendered new frameworks of authority and new forms of political coordination and networking.

The features of governance that have recently concerned geographers include the evolving role of non-state actors and their roles and responsibilities, the alteration of hierarchical boundaries of the state through decentralization and the related emergence of non-governmental organizations and the community/voluntary sector in delivering services. How such shifts are being regulated and presented is increasingly important. Geographers have also debated whether new modes of governance have led to a decline or reassertion of state authority (Shepherd 2002). The nature of new networks of governance and their sustainability has also been studied (Bridge 2002; Yeung 2002). Debates on governance have evolved from wider discussions concerning the future of the nation-state as geographers contend that globalization has meant that the capacity of the state is insignificant to cope with a 'shrinking' world' and the domination of world markets by multinational capital, and too large to provide meaningful service provision in societies that are increasingly fractured by social class, ethnic division and increasingly complex labour markets. According to Jessop:

> Prognoses include the development of an entirely new kind of state; the rescaling of the nation-state's powers upwards, downwards or sideways; a shift from state-based government to network-based governance; or incremental changes in secondary aspects of the nation-state that leave its core intact. (2004: 95)

Evolution and Debate: Contested Governance

There is a series of problems concerning the use of the term governance, especially given that it can be used in a vague manner that does neverthless make appreciable sense within various contexts. It is, for example, often argued that governance aims to end social exclusion through a series of institutional-led fixes. But the reason that many people are excluded is based upon the operation of global economic networks that create variations in wages, working conditions and the refusal of the right to join trade unions. Poverty and social exclusion could also be linked to the uneven nature of wealth distribution but few governments are going to adopt radical policies of governance that redistribute wealth or question the morality and workings of capitalism. In basic terms governance and the applicability of it can be both propaganda led and highly rhetorical (Swyngedouw 1997; Wood et al. 2005). World leaders promote the idea that globalization will promote better governance and more equitable futures for underdeveloped societies. When the US, Canada and Mexico negotiated the North American Free Trade Agreement in the early 1990s, for example, the respective governments told workers and farmers that their livelihoods would not be destroyed, even though most economists acknowledged that many peasants in Mexico and factory workers in the US would lose their jobs. Indeed, critics argued that globalization would provide even greater control for those who dominate the world market (Quigley 1995).

Access to power is also uneven because of the growing dominance of multinational capital and the financial capacity that it holds. This financial clout permits the manipulation of the political system via lobbying and political 'donations' (Jacobs and Shapiro 2000). Increasingly, elected representatives are influenced by financial largesse and this means that big business can directly influence government policy (i.e. the nuclear power industry, the arms industry and transport industries) despite public opinion. Without this more critical appreciation of power it is inaccurate to describe governance as benign public sector management or to view the relationship between government, business and civil society as uncomplicated. In short, while governance is usually related to the term 'good', the complexities of modern societies make this an extremely patronizing and condescending view that masks many of the struggles and conflicts that exist.

Many geographers have also been concerned that alteration in the meaning of borders and territories has been undertaken via modes of governance that aim to encourage the development of neoliberal globalization. (For more information on neoliberalism see Chapter 13.) The shifting balance of power between nation-states, local and federal governments, multinational corporations and supranational institutions (such as the European Union and World Bank) is being modified. A consequence of this changing political landscape is the redistribution of power and responsibilities between governments and citizens. Rapid globalization and more complex spatial divisions of labour are producing connectivities that are displacing established forms of spatial governance and spaces of citizenship. Economies that are deeply pervaded by mobile capital investment lose control over economic and social planning as the influence of foreign direct investment subverts the power of the nation-state. This raises the spectre that the use of governance strategies to advance neoliberalism has profound implications for the capacity of states and supranational bodies to reduce inequality, sustain social justice and deliver a sustainable regime of social regulation.

Peck's (2006) work has recently developed a critical analysis of the intellectual project of the 'new urban right' and has highlighted how the role of conservative and free market think tanks in the formation, elaboration and diffusion of market-oriented or 'neoliberal' urban policy in the period since the late 1970s has undermined welfarist modes of urban governance. He argues:

> These modes have increasingly given way to an alternative – but hardly stable – regime, based on the invasive moral and penal regulation of the poor, together with state-assisted efforts to reclaim the city for business, the middle classes, and the market. But there is a complex geography to this process of ideological diffusion and transformation. (2006: 681)

Boyle and Rogerson (2006: 226) have shown how ideas of governance are increasingly attached to modes of governance that are self-regulating and supposedly located in self-reproducing local communities. As they argue, the idea of governance when applied to deprived communities can 'in fact be thought of as a thinly veiled moral crusade against vulnerable residents'. Moreover they assert that the rhetoric of commitment and inclusion

> might be best conceived as a 'flanking support' for the neoliberal turn in urban governance in British cities; morally commendable communities are defined as those who can reattach themselves to the 'mainstream' and stand on their own two feet within the terms set by neoliberal market economics. When these morally charged interventions fail to connect locally, they have the potential to stir conflict over who has the authority to judge forms of community life. Mapping and accounting for the uneven development of moral conflicts over community is therefore a pressing concern.

Case Study: Global Governance and Loan Conditionality

Supranational entities such as the International Monetary Fund, the World Trade Organization, the World Bank and the Group of 8 (G8) have presented themselves as the promoters of 'good' governance and global stability. Closer analysis would examine the failure of the United Nations to prevent genocide, the incapability of the World Trade Organization to avert the exploitation of defenceless workers and, in the case of the World Bank and International Monetary Fund, to avert increasing global financial catastrophes or significantly reduce poverty in developing countries. Furthermore it could be noted that these organizations of governance are western-centric and tied to an unswerving neoliberal platform (Castells 2001; Chossudovsky 1997; Jones et al. 2004a; Raco and Imrie 2000).

The power of these institutions lies in their ability to influence global governance through the control of capital and the provision of funds to less developed nations. In particular and with regard to promoting 'trade', these institutions promote a form of 'globalization' that is linked to creating open economies and a borderless global economy. This eradication of regulatory trade is presented as the development of a new form of governance that will undermine poverty and spatial unevenness within the global economy. However, such governance aims to deterritorialize through a reconfiguration of geography, so that space is no longer wholly mapped in terms of territorial places but understood as being linked to a spatial organization of social relations and transactions (Giddens, 1990: 4) that is achieved by the promotion of interregional flows and networks of economic activity (Raco and Imrie 2000).

In 2005 the summit of G8 leaders in association with the World Bank and International Monetary Fund set itself UN Millennium Development Goals to be achieved by 2015. These included targets and modes of governance that would eradicate extreme poverty, combat HIV, AIDS and malaria, and ensure that every child receives primary education. Some of the 'goals' of what these leaders promoted as 'good governance' included:

- a doubling of aid by 2010 – an extra $50 billion worldwide and $25 billion for Africa;
- writing off immediately the debts of 18 of the world's poorest countries, most of which are in Africa. This is worth $40 billion now, and as much as $55 billion as more countries qualify;
- a commitment to end all export subsidies;
- the statement that developing countries will 'decide, plan and sequence their economic policies to fit with their own development strategies, for which they should be accountable to their people';
- funding for treatment and bed nets to fight malaria, saving the lives of over 600,000 children every year;
- full funding to totally eradicate polio from the world;
- the goal that by 2015 all children will have access to good quality, free and compulsory education and to basic health care, free where a country chooses to provide it;

However, to enjoy this mode of 'new governance' developing countries had to tackle corruption, boost private sector development and eliminate impediments to private investment, both domestic and foreign (Monbiot, 2005). Some countries would also have to privatize state owned resources and welfare services. These conditionalities are the policies governments must follow before they receive aid and loans and debt relief. As argued by George Monbiot:

> And here we meet the real problem with the G8's conditionalities. They do not stop at pretending to prevent corruption, but intrude into every aspect of sovereign government. When the finance ministers say 'good governance' and 'eliminating impediments to private investment', what they mean is commercialisation, privatisation and the liberalisation of trade and capital flows. And what this means is new opportunities for western money .... There is an obvious conflict of interest in this relationship. The G8 governments claim they want to help poor countries develop and compete successfully. But they have a powerful commercial incentive to ensure that they compete unsuccessfully, and that our companies can grab their public services and obtain their commodities at rock-bottom prices. The conditionalities we impose on the poor nations keep them on a short leash. (2005: 21)

Figure 3.1 Demonstrators march through the intersection of 18th and M Streets NW in Washington DC at a demonstration against the World Bank and Internationa Monitory Fund 16 April 2005

Figure 3.2 Mainstream march, part of the October Rebellion demonstrations against the World Bank and IMF in Washington DC

In the last decade, protesters inside and outside the developing world have challenged the conditionalities that organizations like the IMF put on countries (see Figures 3.1 and 3.2).

It is important that geographers, when considering issues of governance, bear in mind that many governments and supranational bodies see policy and the language of democracy as an instrument that can be utilized as part of the broader continuum of power leverage that they can use to achieve their goals – a case of thinking about a definition of governance that might draw upon the question 'who controls governance, and for whom?'

KEY POINTS

- Governance is important with regard to examining the exercise of power.
- Political and legal authority, the regulation of the market and the use of force are combined to produce social regulation.

- It is assumed that good governance is constructed around heterarchy (inclusion) rather than hierarchy (domination).

NOTE

1 A regime of accumulation and social regulation refers to the type of economy a state supports and the mechanisms by which it regulates the social actors involved in it (Lipietz 1992). Most western states have a capitalist regime of accumulation. The two most important social roles attached to capitalism–workers and owners–are regulated by convention and a legal code to enforce it. Workers and owners accept their roles because the state ensures that both parties gain something from the arrangement. Workers, for example, are guaranteed by law a minimum income, a 40-hour work week with pay for overtime, and benefits. Owners, by extension, are guaranteed reasonable tax rates and the ability to invest their profits as they see fit.

FURTHER READING

Castells, M. (2001) 'Information technology and global capitalism', in W. Hutton and A. Gidders (eds), *On the Edge: Living with Global Capitalism*. London: Vintage. pp. 52–74.

Chossudovsky, M. (1997) *The Globalization of Poverty: Impacts of the IMF and World Bank Reforms*. London: Zed Books.

Monbiot, G. (2005) 'A truckload of nonsense', *The Guardian*, 14 June.

4 DEMOCRACY

Mary Gilmartin

Definition: Better than a Dictatorship, but not always Radical

Definitions of democracy often begin with the origins of the word: in ancient Greece, and as a form of rule by the people. In this way, democracy is contrasted to monarchy, a form of government where authority is vested in a single, hereditary figure. As a concept, democracy is radical. It suggests that people should be directly responsible for governing their own societies. In practice, however, the radical core of democracy is often undermined and the power of the people diminished. This happens in a variety of ways: from the creation of more indirect forms of democracy, to the use of the concept as a justification for new forms of imperialism. Iris Marion Young comments that 'democracy is hard to love' (Young 2000: 16), but recognizes that for ordinary people, democracy offers a means of restraining rulers, influencing policy and promoting justice (Young 2000: 17).

A variety of forms of democracy have been identified. One distinction is between direct democracy and indirect democracy. Peter Taylor has suggested that direct democracy involves rule by the people, while indirect democracy occurs on behalf of the people, usually involving elections with competing political parties (Taylor 1993: 328). Other forms of democracy include formal, participatory, social and liberal. O'Loughlin argues that four key features – free and fair elections, universal suffrage, the accountability of the state apparatus, and freedom of expression and association – characterize a formal democracy. When there is a high level of democratic particpation without differences across social categories, this is defined as participatory democracy. Liberal democracies include the features of formal democracies, but also involve civilian control over the military as well as protection for individual and group freedoms (O'Loughlin 2004: 27). A fourth distinction is social democracy. Taylor suggests that under social democracy, states take responsibility for the welfare of their citizens, fund their welfare through taxation and work within a political consensus that significant welfare expenditure is necessary and just (Taylor 1993: 249). However, recent political geography texts pay scant attention to social democracy as a concept and focus instead on liberal democracy. It is also important to highlight the concept of transnational democracies – democracies that extend beyond state borders, for example through the work of transnational civil society.

Evolution and Debate: Three Ways to View Democracy

There are many ways in which political geographers have engaged with the concept of democracy. The first way takes democracy as a given and looks at the ways in which democracy operates. This has traditionally taken the form of a strong interest in voting and elections and is often called electoral geography. The second way is concerned with democracy as a process, with a particular focus on processes of democratization. This has gathered momentum in recent years, particularly with the end of the Cold War and the break-up of the Soviet Union. The third approach is often described as critical and focuses on providing critiques of democracy, particularly as it is practised, and on providing alternative definitions and understandings of democracy.

Electoral Geographies

Taylor has suggested that electoral geography was concerned with the geographies of voting, with geographical influences on voting and with the geographies of representation (1993: 230–40). Attempts to understand the geographies of voting initially concentrated on spatial voting patterns, seeking to explain the patterns and later attempting to understand those patterns through broader explanatory frameworks. Research on the geographical influences on voting focused on local influences on voting behaviour. This included the identification of the 'friends and neighbours' effect, which suggested people were more likely to vote for local candidates, as well as the neighbourhood effect, which attempts to explain why parties perform better than expected in their strong areas. In these ways, political geographers have applied quantitative spatial analysis to the study of voting, at local and regional as well as national levels.

In addition to the study of voting patterns, political geographers have also attempted to address broader questions of representation and fairness in the practice of democracy. Two particular areas of interest were gerrymandering and malapportionment. Gerrymandering refers to the drawing of electoral district boundaries to favour one group (often racially or ethnically defined) over another; while malapportionment refers to the creation of electoral districts of an unequal voting population size (see Johnston 2002 for a discussion of both in the UK). Richard Morrill has suggested that the quality of representation in western democracies is currently under scrutiny (Morrill 2004). In relation to the US, he identifies a number of key problems, including the representation of minorities, disenfranchisement, the displacement of power to more remote authorities, and corruption (Morrill 2004: 84–89). His observations have much broader relevance and could equally be applied to a number of western European democracies such as Germany, France and the UK.

Democratization

The question of democratization is a second, more recent area of concern in political geography. Writing in 1991, Huntington suggested that there have been three waves of democratization (Huntington 1991). The first began in the US in the nineteenth century and continued until 1922, involving the spread of parliamentary democracy to Europe, North America, Australia, New Zealand and parts of Latin America. The second began after World War II and applied to some postcolonial states and part of Europe. The third began in the early 1970s and continues today, though its key point was in the early 1990s with democratization in Central and Eastern Europe (Jones et al., 2004b: 138). This recent 'wave' of democratization has received attention from some political geographers, who have examined processes of democratization in Eastern Europe and the former Soviet Union. This is discussed in more detail below.

Within political geography, the study of democratization has also been expanded. In particular, democratization has been examined not just in terms of formal voting rights, but also in terms of the role of civil society. Such studies often focus on southern hemisphere countries and regions, particularly in relation to questions of 'development'. For example, Juanita Sundberg's study of Guatemala shows the relationship between processes of democratization and the country's environmental movement (Sundberg 2003). She concludes that new social movements offer the most promising signs of democratization in Latin America (Sundberg 2003: 737). In contrast, Claire Mercer has pointed out that western academics often assume that strengthening non-governmental organizations (NGOs), and thus civil society, in the developing world will lead to greater democratization (Mercer 2002). Mercer suggests that this assumption is problematic, and argues that NGOs play a more complex role in processes of democratization than is often acknowledged. These approaches address concerns raised by Bell and Staeheli (2001), when they highlighted the difficulties of assessments of democratization that use measures that are purportedly universal, but that fail to take local or regional specificities into account.

Critical Approaches

Third, more recent studies of democracy within political geography have adopted approaches to the concept that are described by John Agnew as 'self-consciously critical perspectives' (Agnew 2002: 173). One such perspective is outlined by Barnett and Low in the introduction to their edited (2004) text on spaces of democracy. They comment that while the most significant recent global trend has been the spread of democracy, it has had little effect on research agendas in human geography (2004: 1). They suggest that the focus on how democracy works elides a more normative engagement with the concept of democracy. In particular, they comment that 'the universalization of democracy as a taken-for-granted good does not imply that the meaning of democracy is cut and dried' (2004: 16), and suggest that their edited text opens up discussion and debate about alternative understandings of democracy. More recently, Ettlinger has provided one alternative understanding of democracy in practice in post-Katrina New Orleans (Ettlinger 2007). Drawing on the work of Iris Young, Ettlinger argues that it is imperative that a rebuilt New Orleans embeds 'principles of inclusive, everyday democracy' (Ettlinger 2007: 11), paying attention to social and economic mixing and interaction and to cooperative rather than confrontational politics.

Case Studies: A Sampling from the US, Eastern Europe and Mexico

Given the three broad areas of focus for political geographers interested in questions of democracy, this section provides examples of recent work in each of the areas. The discussion of electoral geographers looks at the analysis of US presidential elections in 2000 and 2004. The discussion of democratization focuses on ways to measure and assess the process. Finally, the discussion of critical approaches to democracy concentrates on alternative spaces of democracy, beyond voting in state-sponsored elections.

Electoral Geographies: Anomalous Patterns in the 2000 and 2004 US Elections

According to Jones et al. 'there is a dearth of good up-to-date literature on electoral geography' (2004b: 156), an assertion that is repeated by Morrill, Knopp and Brown (2007) in relation to the US Presidential elections in 2000 and 2004 when they comment that most studies of the elections were written by non-geographers. In many ways, their article is an attempt to reassert the importance of electoral geography, and to insist on the centrality of geographical contexts for understanding the ways in which democracy works. Morrill et al. carried out a detailed quantitative analysis of voting patterns at a county level. They were particularly interested in counties they described as 'anomalous', where 'local-scale results ... contradict conventional wisdoms and dominant analyses of election results' (Morrill et al. 2007: 526). Among the anomalous patterns they identify are large metropolitan areas and counties with a predominance of racial minority voters who voted Republican, as well as non-metropolitan areas that voted Democrat. Morrill et al. also analysed changes in voting patterns from 2000 to 2004 (see Figure 4.1). They tentatively suggest that a reduction in the Republican margin in some counties may be linked to a reaction against religious fundamentalism, while an increase in the Democratic margin in some counties that are narrowly Democratic may be linked to progressive religious and political traditions (Morrill et al. 2007: 548). However, they also admit that their hypotheses are questionable and that qualitative analysis will be necessary to test them. In a conclusion that points to the possibility of a new direction for electoral geography, they claim that a 'rich, mixed-methods approach' may provide 'entirely new and more

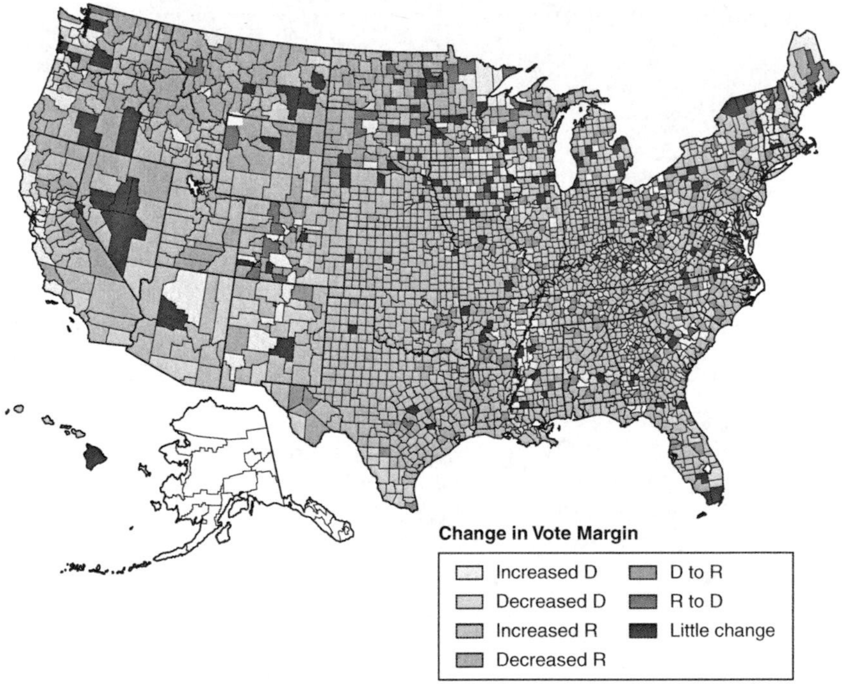

Figure 4.1 Change in voting margins, US President 2000 and 2004.
(Reproduced with permission)

nuanced interpretations' of voting patterns and electoral geographies (Morrill et al. 2007: 549).

Democratization: Consolidation in Eastern Europe

One of the key writers on processes of democratization is John O'Loughlin. His survey article on the topic, outlining the ways in which the global spread of democracy may be measured and explained, takes the Freedom House measures as a basis for assessment (O'Loughlin 2004). Freedom House carries out yearly surveys of the political and civil rights and freedoms of all countries. On the basis of these surveys in 2001, O'Loughlin argues that the clearest evidence of democratization was in some previously communist countries in Eastern Europe and the former Soviet Union. In this way, he suggests, the 1990s displays evidence of a regional character to democratization, rather than the more global 'Third Wave' of democratization suggested as inevitable by commentators such as Samuel Huntington (O'Loughlin 2004). A more detailed study of Kyrgystan (Kuchukeeva and O'Loughlin 2003) suggests that democratization involves the consolidation of eight key institutional elements: freedom of assocation, freedom of expression, the right to vote, eligibility for public office, the right to compete for support/votes, free and fair elections, alternative sources of information and the dependence of government policy-making on votes (2003: 557). The authors conclude that civil society – in Kyrgystan and elsewhere – has a crucial role to play in the process of democratization. More recently, Lemon's study of democratization in southern Africa (Lemon 2007) suggests that the presence of dominant political parties in many countries creates difficulties for minorities and also warns that the equation of democracy with socio-economic benefits may lead to disillusionment with the process. For Lemon, true democracy should be internally generated rather than externally imposed and should address divisions in society without further institutionalizing them (2007: 847).

Critical Approaches: on the Fringes in Chiapas, Mexico

The issue of social movements, particularly transnational social movements, has been a key focus of critical political geographers interested in the ways in which democracy might be reimagined. John Agnew has argued that 'the ability of transnational networks to offer an alternative to conventional territorial forms of representation is open to more than reasonable doubt' (Agnew 2002: 167). However, other commentators assert that transnational social movements 'give voice to people and causes outside the established power structure' (Miller 2004: 223) and, in doing so, help to transform official institutions of democracy. One such social movement is that of the EZLN (Zapatistas) in Chiapas in Mexico. The Zapatistas came to prominence on 1 January 1994, the date NAFTA (the North American Free Trade Agreement) came into effect. With between 5,000 and 15,000 rebels (one third of whom were women), the EZLN occupied San Cristobal de las Casas, the capital city of Chiapas, and other towns around the state (Froehling 1999: 166). The purpose of the uprising was to draw attention to poverty and inequality in Chiapas, but also to link processes and outcomes in Chiapas to those in other parts of the world. '"This isn't about Chiapas – it's about NAFTA and [President Carlos] Salinas's whole neoliberal project' explained the EZLN's chief spokesperson, Sub-comandante

Figure 4.2 Zapatista Primary School

Marcos on the first day of the uprising' (in Jeffries 2001: 129). In other words, the Zapatistas used their uprising to declare war on neoliberal globalization. This struggle continues (see Figure 4.2).

The Zapatistas have gained widespread attention from political geographers interested in critical approaches. One area of interest is the Zapatistas' call for indigenous autonomy within the Mexican state (Gallaher and Froehling 2002). Rather than calling for separation, the Zapatistas want to remain within the state. However, they argue that for this to happen, indigenous communities have to be equal to the federal government. This equality will only be possible if indigenous communities are given 'the space and the freedom to rebuild collective forms of decision making, governance and economic subsistence integral to indigenous ways of life' (Gallaher and Froehling 2002: 93). A second area of interest is the way in which the Zapatista movement has given rise to other global networks and social movements, such as the People's Global Alliance, which is interested in further internationalizing social movements (Routledge 2003a). Routledge suggests that such networks offer alternatives to the 'exploitation and domination of the new world order' (2003a: 246), thus providing an counter-hegemonic and transnational understanding of democracy in practice.

KEY POINTS

- In theory, the concept of democracy suggests that people should be directly responsible for governing their own societies. In practice, the power of people is often diminished by the creation of indirect forms of democracy.
- Political geographers have paid particular attention to how democracy works, both through detailed studies of voting and elections and, in recent years, through examining the process of democratization.
- In recent years, political geographers have also started to argue for the recognition of alternative spaces of democracy, such as transnational social movements.

FURTHER READING

Barnett, C. and Low, M. (eds) (2004) *Spaces of Democracy*. London: Sage.

Johnston, R. (2002) 'Manipulating maps and winning elections: measuring the impact of malapportionment and gerrymandering', *Political Geography* 21(1): 1–31.

Morrill, R., Knopp, L. and Brown, M. (2007) 'Anomalies in red and blue: exceptionalism in American electoral geography', *Political Geography*, 26(5): 525–553.

Young, I.M. (2000) *Inclusion and Democracy*. Oxford: Oxford University Press.

Part II

Modes of Power

INTRODUCTION

Carolyn Gallaher

Power is inescapable. It affects everyone and its presence can be felt in ways big and small. When a prisoner of war is beaten by his or her captor, power is felt on the body. The pain the prisoner feels signifies a distinct hierarchy between those who hold power and those who do not. Power can also operate more subtly. When the child of an African refugee tries to find a job in France, she may find it difficult because of discrimination. She will not acquire any bruises or physical injuries in her search, but power will structure her relationship with potential employees as much as it does a prisoner with his captors.

When political geographers examine power, they look at its spatial manifestation. In particular, they ask where power is concentrated and why. Political geographers tend to analyse the distribution of power among states, but increasingly they also analyse how power is embodied individually and by distinct groups.

This part of the book begins with the concept of **hegemony**, a form of power that is based not only on raw power (e.g. the monopolization of the means of violence), but the consent of those governed. The concept of hegemony helps geographers explain how states justify their decisions, even unpopular ones, so that society follows established rules and norms even when they go against certain members' interests.

The second concept covered is **territory**. Geographers use the concept of territory to describe the spatial basis of power. They examine how states, social movements and ethnic/religious groups use territory to ground their power in space. This process often entails creating exclusionary spaces. For the in-group, these spaces are designed to feel safe. For those in the 'out-group', such spaces can feel unwelcoming, even dangerous.

The third concept, **geopolitics**, is in many ways territory writ large; geographers examine the balance of power between states with reference to their control or influence over strategic spaces, such as oil deposits, water resources, etc. As the introduction notes, geopolitics has a chequered history in that it was often used to justify exploitation during the colonial period. However, today critical geopolitics examines the spatial balance of power with an emphasis on how power relations establish inequalities in the world system. Rather than working with foreign policy establishments, as in older eras, scholars who follow critical geopolitics concern themselves with social justice, human rights, economic justice and the environment

In the final chapter we consider a particular type of geopolitical power arrangement – the **superpower**. In the world system superpowers can have competitors, as the US and Soviet Union did between 1945 and 1989, or they can exist as the sole dominant power, as some suggest the US is today.

To bring these concepts alive, case examples are provided from Afghanistan, the United States and Iraq, among many others. As these examples demonstrate, the constitution of power is inherently spatial. Indeed, while the powerful can exercise their power across wide swaths of territory, local customs and systems of meaning shape how it is felt and experienced.

5 HEGEMONY

Carolyn Gallaher

Definition: By Consent as Much as Coercion

The term 'hegemony' is a concept used to describe the processes by which people are ruled or governed. The singular form of the term, hegemon, can refer to an individual leader, a state or a collection of states acting together. While hegemony is often used interchangeably with terms like power and domination, its meaning is more specific (Agnew 2003b). Hegemony describes a form of power gained at least in part through consent. The study of hegemony, then, tends to focus not only on the coercive elements of power, but also on its naturalization and legitimization. Contemporary understandings of hegemony stem from the work of Antonio Gramsci, an Italian Marxist who developed a theory in the 1930s to explain why the working-class revolution predicted by Marx had failed to materialize (Figure 5.1). While the concept is

Figure 5.1 Antonio Gramsci, around 30 years old, c. 1920

generally associated with Marxism, it is increasingly used, in an adapted form, by poststructuralists. Geographers use the concept in both ways.

Gramsci and His *Prison Notebooks*

In 1926 Antonio Gramsci was imprisoned by Italy's fascist Prime Minister Benito Mussolini. As founder of the Italian communist party and a vocal critic of the regime, Gramsci was deemed a threat to Mussolini's then budding regime (Fiori 1996). In prison Gramsci was given access to pen and paper and he used them to good effect, filling dozens of notebooks with political meditations. His notebooks were published posthumously in a volume entitled *The Prison Notebooks*. In them, Gramsci tried to tease out why Italy's political system was more autocratic than those of other states in Europe. He used the notion of hegemony to answer this question (Adamson 1980).

Gramsci theorized that Italy's political culture varied from most of Europe because the Italian bourgeoisie had never attained hegemony, which he defined as a state of power based on the consent of those governed or ruled. In other European states, the bourgeoisie class had seized power and was able to depict its class interests as national interests. This in turn allowed the bourgeoisie to rule with the consent of those with different class interests. Gramsci theorized that the absence of such a hegemonic regime in Italy permitted a coercive regime like Mussolini's fascist government to take root (Gramsci 1971).

Gramsci also used the notion of hegemony to explain why workers had failed to overthrow capitalism in Europe's industrial centres. He argued that the working class needed to establish the hegemony of socialism as an ideal before it could take over the state apparatus and destroy capitalism. He referred to this process as a 'war of position' because workers would need to develop support for communism both within *and* beyond the working class. Once consent was established, working-class leaders would be able to institute a 'war of movement' in which the capitalist state is overthrown for a communist one (Gramsci 1971).

Gramsci's theory of hegemony represented an important break in Marxist thinking at the time. His focus on consent as an integral part of capitalist governance pushed Marxism out of its increasingly determinist corner by drawing attention to the contingencies of the social world not considered or emphasized by Marx (Taylor 1996).

Poststructuralism and the Concept of Hegemony

In the mid 1980s poststructuralists adapted Gramsci's concept of hegemony, building on the constructivist elements of Gramsci's theory while rejecting elements considered reductionist (Smith 1998). Ernesto Laclau and Chantal Mouffe's (1985) work on hegemony is a case in point. In contrast to Gramsci, and other Marxists who located hegemony in the ability of the bourgeoisie to naturalize its interests as those of the wider polity, Laclau and Mouffe argue that hegemony can also develop around the interests of non-class-based social groupings, such as ethnic groups. Indeed, in the West, white interests have long been normalized as those of the polity as a whole, even though minority interests are often at odds with those of the white majority. For Laclau and Mouffe, hegemony is better understood as the ability to fix meaning, through language, around so-called nodal points of identification. In this configuration, certain social categories, of race, gender and sexuality (among others), become normative or hegemonic while their others are considered 'different,' 'inferior' and, potentially, 'dangerous'. One can, therefore, speak of white male hegemony. In the fashion world, for example, whiteness is often held up as the standard by which individual beauty is defined. People with brown or black skin are considered less beautiful than whites, or even unattractive. Likewise, the assumption that males are better at science makes it difficult for women to get good jobs in the sciences, even when they have excellent qualifications.

Laclau and Mouffe also take issue with the Marxist political project, of which Gramsci's 'war of position' and 'movement' broadly fit. In particular, they identify two shortcomings. The first critique was to be found in the totalitarian excesses of actual Marxist regimes. The brutality of Stalinism in the Soviet Union, for example, and Pol Pot in Cambodia, was impossible to ignore, and wholly indefensible. Indeed, millions of innocent citizens were killed by both regimes. Second, Marxist theory was unable to explain the rise of so-called new social movements, such as feminism, black power and gay rights. The strict adherence to class tension as the driving impetus behind social upheaval left Marxists flat-footed and unprepared to offer nuanced explanations for the rise of these movements in the 1960s. Many scholars also sympathized with these movements and were loath to write them off, in classic Marxist fashion, as victims of false consciousness.

Central to Laclau and Mouffe's configuration of radical democracy is the idea that differentiation is a fundamental part of social life. Individuals will always define themselves in opposition to others and this will lead, at the macro level, to group identity formation. This view puts Laclau and Mouffe at odds with both liberal viewpoints that call for the erasure of difference (i.e. invocations to be colour blind) and Marxist ones that label social activism outside of class politics as false consciousness. For Laclau and Mouffe a radical democratic politics must focus on renegotiating the terms by which social differentiation occurs. The problem is not difference, but difference that is cast antagonistically, where the other is viewed as dangerous and worthy of destruction. (For more information on the other see Chapter 28.) A radical democratic politics must encourage agonism, where groups recognize their differences and may even dislike each other, but refuse at the same time to view difference as illegitimate (Mouffe 1995). In an effort to combat right-wing hegemony, Laclau and Mouffe call on the left to build a chain of equivalencies between class-based politics and newer identity-based movements on the left.

Evolution and Debate: Hegemony through Different Lenses

Political geographers analyse hegemony in one of two ways. The traditional approach examines hegemony as it pertains to world orders and their territorial manifestations (Agnew and Corbridge 1995). Some scholars working in this vein emphasize the role of political economy in the constitution of world orders (Agnew and Corbridge 1995). Others stress the constructed nature of world orders, focusing on the discursive realm in which alliances are legitimized and threats are naturalized (Flint 2001). Some scholars combine the two (Peet 2002). In all of these studies, however, the focus is on macro-level analyses, where states or collections of states are the primary actors. A more recent approach, influenced by poststructuralism generally and feminism more specifically, examines hegemony in the context of identity politics. This approach focuses on the construction of hegemonic categories of race, gender and sexuality, and the spatial manifestations thereof (Hyndman 2004a; Sharp 2000). Attention is also given to the construction of others within a hegemonic system of identity and the social and spatial marginalization they experience. Feminists have tended to focus on marginalization at subnational scales – regions, cities, and neighborhoods. However, a number of studies have also examined how marginalization varies across different scales. Within political geography these approaches have tended to operate in isolation from one another. However, recent scholarship suggests there are important avenues for cross-fertilization (Hyndman 2004a, 2007; Steinberg 2003).

Traditional Approaches

For traditional geopolitical approaches the Peace of Westphalia is considered a watershed event. The treaty, which ended the Thirty Years War in Europe in 1648, paved the way for the creation of the modern nation-state system by allowing hegemons to connect their power to discrete and definable territorial units for the first time in history. As the system was concretized and formalized, it also paved the way for inter-state struggles for power. The state or alliances of states that come to dominate the world system develop what political geographers call global hegemony – a set of rules other states follow, largely by consent. However, Agnew and Corbridge reject the 'great power' view of hegemony common in political science which demarcates hegemonic periods by the state(s) most dominant in the world system. In contrast, they argue that global hegemony is at root a spatial phenomenon. It is given form along two geographic dimensions: the 'scale of economic accumulation' and the 'space of political regulation' (1995: 17). Economic accumulation tends to be *territorial*, 'based around tightly integrated national economies', or *interactional*, with trade and investment outside the nation encouraged. Political regulation tends to form around three types of states – *national*, *imperial* or *international*. A national state confines its sphere of influence to the area inside its borders. An imperial one seeks to extend control over space outside its boundaries. An international state manages affairs outside its boundaries to secure its accumulation, but may or may not control those spaces directly.

The spatial basis of hegemony also has relevance for our understanding of the rise and fall of global hegemons. While traditional explanations put the rise and fall of global hegemons down to the shrewdness a hegemon employs *vis-à-vis* potential competitors, geographers see the rise and fall as related to economic and technological changes that favour one spatial organization over another.

Agnew and Corbridge (1995) identify three periods of global hegemony since the Peace of Westphalia. The first, the 'British Geopolitical Order', extended from roughly 1815 to 1875. During this period, the British colonial state was the dominant global force. Its hegemony was based on accumulation of British colonial holdings and the international scope of these forced it to operate as an international state. The second period of global hegemony was marked by imperial rivalries between the United States, powerful European states and Japan. Accumulation was primarily organized territorially, in the 'home' country, or in areas each power sought to control to the exclusion of the others. The enmity this created helped foment two world wars. The final phase pitted two imperial states – the US and the Soviet Union – against one another in an effort to control accumulation at a global scale.

Political geographers are less certain as to how to characterize the current stage of global hegemony, although most scholars argue that the United States is important to it. In the case study section I describe competing views of American hegemony.

Hegemony and Identity Politics

Traditional conceptions of hegemony have come under increasing scrutiny, particularly by feminist geographers, who argue that traditional views focus too closely on the role of the state and its interactions with other states (Hyndman 2004a; Sharp 2000). In so doing, traditional views have failed to examine how global hegemonies are embodied differently across social categories and space. A single-minded focus on the state also privileges the 'public' over the 'private.' As Joanne Sharp (2000: 30) notes:

> Geopolitics thus posits a separation of the sphere of international politics from a sphere of domesticity and the politics of day-to-day life. This division perpetuates privilege by implying that international politics is beyond the ability of most people, and thus should be left to experts.

As an antidote, feminists and others examine hegemony as a process by which certain identities are afforded dominance, and through which their particular viewpoints and related concerns come to be cast as universal. Joanne Sharp's analysis of depictions of the Cold War in the American magazine *Reader's Digest* provides a case in point. A decidedly middlebrow publication, *Reader's Digest* mixes advice for the lovelorn with discussions of American foreign policy. Its readership, when Sharp published her book, was roughly 16 million Americans. Though respectable, the magazine would never be considered a definitive source of American foreign policy. Sharp argues, however, that the magazine's location in the realm of popular culture makes it a fruitful place from which to analyse American hegemony during the Cold War. As she notes, 'geopolitical knowledge does not simply "trickle down" from elite texts to popular ones'. The production of geopolitical knowledge is intertextual – a two-way process – so *Reader's Digest* is as important a site of analysis as presidential papers. Also, because *Reader's Digest* was never seen as explicitly political, most readers accepted its analysis with less suspicion than more overtly political speech. Reading *Reader's Digest* as a hegemonic text can, therefore, refocus our attention on the consensual aspects of hegemony – the 'why people accept what they are told' part of the equation (Sharp 2000: 36).

Case Study: Is the United States a World Hegemon?

When the Berlin Wall fell in 1989, many in the West celebrated the triumph of capitalism over communism (see Figure 5.2). The political scientist Francis Fukuyama (1989) famously declared 'the end of history', arguing that humankind had finally reached its pinnacle of political and economic organization and that peace and stability would spread across the globe. However, the decades after the Wall fell proved to be far more turbulent than Fukuyama predicted. New wars emerged in Africa and Europe; the transition from communism unfolded chaotically; and financial crises rocked Asia. And, on 9/11, 2001 Al Qaeda, a transnational terrorist group, flew planes into the World Trade Center and the Pentagon, destroying or severely damaging the most important elements of the American military industrial complex's symbolic repertoire.

Like their peers in other disciplines, political geographers have grappled with identifying the key features of this new era. There is not yet

Figure 5.2 Remnants of the Berlin Wall. After the Berlin Wall fell in 1989 some scholars argued hegemony was polycentric rather than centred in one state

a consensus on what this new era will be defined by, or what rules will govern it. Most scholars agree, however, that whatever form contemporary hegemony takes, American interests are crucial to it. Below are two vignettes to describe how different political geographers have theorized contemporary hegemony and America's place within it.

Hegemony without a Hegemon?

In *Mastering Space* (1995), which was written after the fall of the Berlin Wall, but before the attacks of 9/11, 2001, John Agnew and Stuart Corbridge label the current hegemonic regime *transnational liberalism*. Economic accumulation in the contemporary regime is *interactional* and relies on internationalized states to regulate its territorial imperatives. Agnew and Corbridge posit that no one state dominates the contemporary period; the system is polycentric.

Transnational liberalism works primarily through consent. It has its roots in *Pax Americana*, the previous hegemonic regime formulated in the aftermath of World War II. The goal of *Pax Americana* was to create a world economic system based on free trade. The muscle behind it was established in 1944 by the Bretton Woods Agreement, which laid out an organizational apparatus to encourage *Pax Americana*'s goal. The General Agreement on Tariffs and Trade (GATT) and the World Trade Organization (WTO) would establish norms of free trade and procedures for recourse against protectionism while the World Bank and the International Monetary Fund (IMF) would ensure a steady cash flow for anti-communist development. States were, however, central to the system. Indeed, organizations like the World Bank and the IMF could only work with state governments, not individuals, families or other sub-national groupings.

Over time, however, accumulation strategies encouraged by *Pax Americana* weakened state power relative to that of corporations and finance capitalists (Agnew and Corbridge 1995; George 1990). In particular, the 1980s debt crisis hastened the shift from *Pax Americana* to transnational liberalism. During the crisis, states in Latin America, Africa and Asia were forced to take out so-called structural adjustment programmes in order to avoid default. These loan packages required states to 'denationalize' their economies by encouraging foreign investment, privatizing state-owned industries and export-orienting their economies (Naím 2000). For example, African state bureaucracies, which at one time employed up to a third of the working-age population, were systematically dismantled during the 1990s (Ould-Mey 1996). Western states were also affected by transnational liberalism, albeit less intensely. Attempts to dismantle welfare entitlements in the US and western Europe are emblematic of this logic as well (Harvey 2000). So, too, is the drive to outsource industrial and service sector work to areas with cheaper labour.

By the time the Berlin Wall fell in 1989, transnational neoliberalism was in full swing. Agnew and Corbridge speculated that the new era would be marked by a variety of trends. Workers, who lack the same mobility as capital in the new era, would face increasing instability. Some areas or regions would be unable to fully integrate into the new hegemonic system and would over time become anarchic. And, with no one power centre or hegemon, these areas would be ignored or only loosely contained.

Writing in the wake of 9/11 John Agnew addressed what the US military effort might mean for his and Corbridge's notion of hegemony without a hegemon. He argued that American aggression 'can be seen as an attempt to reestablish the US as central to contemporary hegemony by using the one resource – military power – in which the US is still supreme' (Agnew 2003b: 879).

America the Colonial

Another political geographer who has examined America's new role in the contemporary system is Derek Gregory (2004). Gregory's book *The Colonial Present* differs from Agnew and Corbridge's in a number of ways. Writing after the attacks of 9/11, Gregory is necessarily interested in the American response in Afghanistan and Iraq. For Gregory, the contemporary landscape since 9/11 is best described as colonial. And, unlike Agnew and Corbridge's polycentric hegemony, Gregory sees a clear hegemon at its centre. Indeed, he contends that the American 'War on Terror' is 'one of the central modalities through which the colonial present is articulated' (2004: 13). Gregory's approach is also more discursive than that of Agnew and Corbridge. While Gregory acknowledges the importance of political economy to hegemony, his focus remains on the discursive realm in which America has justified its actions since 9/11. The 'War on Terror', he argues, is more than the economic exploitation and political regulation often associated with colonial ventures. The 'War on Terror' 'is an attempt to establish a new global narrative in which the power to narrate is vested in a particular constellation of power and knowledge within the United States of America' (2004: 16).

The way the American government handles the deaths of Afghan civilians killed by American 'smart bombs' says a lot about the hegemonic discourses underpinning the so-called 'War on Terror'. The military has issued few statements about civilian casualties in Afghanistan, even though there have been plenty to document and explain. And, in October 2001, the US government purchased the rights to all commercial images produced by private companies doing satellite photography in Afghanistan. It then restricted access to those images by entities outside of government. The ostensible goal was to protect national security, but Gregory notes that it also prevented journalists from investigating photographs for mass graves, infrastructure destruction and the like.

This callous attitude is out of keeping with both American and international norms from other wars. Gregory argues that this behaviour was legitimized by the 'War on Terror's' discursive foundations – what he labels, borrowing from Michael Shapiro, an 'architecture of enmity'. This architecture was built on an antagonistic binary that pits a 'civilized' West, led by the US against an undifferentiated Islamic world marked by fanaticism and barbarity. This broad architecture allowed the US to lump together, and ignore the differences between, the Taliban, Al Qaeda, journalists, refugees and ordinary civilians.

The US treatment of prisoners of war followed a similar logic. Although the US is a signatory of the Geneva Conventions, which lay out precise rules regarding the treatment of prisoners from state armies, when American forces captured Taliban fighters, who were attached to the Afghan state, it designated them 'unlawful combatants'. The term put the fighters 'beyond the scope of the law' (Gregory 2004: 65) and allowed the US to imprison them in so-called 'extra-judicial spaces' – in Guantanamo Bay, specifically, and a legal limbo more broadly, in which the rules of confinement, charge and recourse are arbitrary and capricious. It was as if these prisoners failed to exist, given that normal methods for identifying, tracking and communicating with prisoners were never established.

How the new hegemonic order will eventually settle is uncertain. What is certain, from a geographic perspective, is the impetus to view hegemony from a geographical perspective, for without it we could not understand the contemporary world where space seems to matter less even as it matters more.

KEY POINTS

- Hegemony describes a form of power gained at least in part through consent.
- In political geography traditional approaches to hegemony examine how powerful states develop and maintain global dominance. New approaches examine how social identities (of race, gender, etc.) become dominant and how their particular interests are defined and accepted as universal interests.
- Some political geographers believe there is no global hegemon in the world today. Other scholars believe the US is a global hegemon.

FURTHER READING

Agnew, J. and Corbridge, S. (1995) *Mastering Space: Hegemony, Territory and International Political Economy*. London: Routledge.
Gramsci, A. (1971) *Selections from the Prison Notebooks of Antonio Gramsci*, edited and translated from the Italian by Quintin Hoare and Geoffrey Smith. New York: International Publishers.
Laclau, E. and Mouffe, C. (1985) *Hegemony and Socialist Strategy: Towards a Radical Democratic Politics*. London: Verso.

6 TERRITORY

Carl T. Dahlman

Definition: The Spatial Extent of the State

Territory is commonly understood as a portion of the earth's surface claimed by or associated with a particular group or political entity. The term 'territory' comes from Latin and the original meaning referred to land located around or belonging to city-states, land that supplied basic resources for urban-dwellers in the classical world (Parker 2004). Territory was important to early political thought. Plato thought it should provide no more than the necessary material conditions for a city-state's inhabitants, while Aristotle identified it as the third element, along with population and government, in achieving security and wealth. The Hellenistic and Roman Empires that eclipsed the city-states, however, did not rest on the idea of territory as a basis for political power, nor did the later medieval states, since the right to rule came from sovereign power embodied in a ruler rather than a territory (Gottmann 1973). All of this began to change during the 1500s and by the time of the Treaty of Westphalia (1648) territory had come to serve as a way of spatially delimiting sovereignty, putatively limiting states to their boundaries for the greater peace (Agnew 2003c; Krasner 1999; Teschke 2003). In modern history, territory typically refers to the spatial extent of state power as well as the material resources it provides. The historical and explanatory veracity of this assumption requires closer scrutiny, however, as regional and global events necessarily challenge our view of politics as belonging to discretely bounded entities.

Despite disagreements over its development, the contemporary organization of sovereign states rests on the definition of state territory found in international law. In short, state territory is defined as the land claimed by the state and recognized as such by others, including the land's subsoil, its internal and adjacent waters, the airspace over these areas, and certain near-shore marine resources (Brownlie 1998: 105–124). The UN Convention on the Law of the Sea provides an important definition of territorial waters – those that lie between a state's shoreline and the high seas – although this is confounded by competing legal ideas about state territory extending to continental shelves. Outer space, by contrast, is widely accepted as a common area – non-exclusive territory – although it is unclear what ceiling exists between a state's airspace and outer space. Although these definitions of territory are embedded in the legal concept of sovereignty, it is important to distinguish the concept of territory as an area from the concept of sovereignty as a modern institution of authority. (For more information on sovereignty see Chapter 2.)

Like the state system it is meant to describe, the concept of the territorial state developed slowly, and largely as a result of Europe's experiences in state-building and empire. The legacies of this history inform two important legal notions regarding territory outside sovereign states: *res nullius* and *res communis*. The former relates to territory that is unclaimed, at least not by any entity recognized as having sovereign status. Unfortunately for colonized peoples, this meant that European powers had created a way to ignore indigenous land use and take territory (Manzo 1996). The latter refers to areas such as the high seas and outer space that belong to the community of states or else to a common humanity. As the rest of the world map has been 'filled in' with mutually exclusive state claims, territory has become a zero-sum game.[1] The most dramatic ways that states may add territory is through conquest, as in Armenia's effort to wrest the Nagorno-Karabakh area away from Azerbaijan (de Waal 2003). International law nonetheless provides a variety of ways in which territory may be exchanged or added without conflict. Treaties of cession, for example, are made between states to transfer territory. More common but less significant are natural processes of accretion and erosion that change shorelines and affect near-shore territorial delimitations. Similarly, avulsion – changes in a watercourse – may add or remove territory where boundaries are defined by rivers, potentially leading to disputes over the delimitation of state territory.

These means of transferring territory between states must be considered separately from territorial disputes resulting from problems during decolonization, as in Kashmir where uncertainty over the line dividing Pakistan, India and China has led to conflict for the last sixty years. Furthermore, the twentieth century has seen an ever-dwindling list of former colonial, mandate, trust and non-self-governing territories, whose status has been resolved through consolidation with existing states or else by decolonization. The United Nations has attempted to mediate and resolve the status of these territories amid competing state claims, indigenous rights movements and frequent violence, which has both slowed the process, as in the Western Sahara where Morocco has tried to stall a referendum on independence. The role of the UN in other disputes has also led to the creation of new states, such as East Timor. Many current territorial issues relate to efforts to break apart states. Partition provides one of the clearest, if also controversial, acts for dividing territory into two or more separate units. Partition typically leads to the formation of at least one new independent state, although parties with irredentist claims may subsequently seek to annex divided territory to an existing state. Imperial powers, particularly those retreating from a colony's internal political disputes, have most frequently employed this rather dramatic reorganization of the world map. Britain's so-called 'divide and quit' policy was employed as it withdrew from the Indian subcontinent by creating two independent states: India as a largely Hindu state, and Pakistan as a largely Muslim state. Temporary partitions have also produced complex and interesting forms of territorial organization. The use of the 17th parallel in the 1954 Geneva Accords that created North and South Vietnam, for example, was meant to be temporary until new elections could produce a unified government. Instead, the division became the basis for separate political developments leading to further war.

Evolution and Debate: Complicating Territory

Geographers have identified a number of problems with the way we understand territory and its role in political life, not to mention its importance for social and economic relations. Three of these problems have spurred useful ways for reconceptualizing territory: first, territory is too often approached as a static object, a backdrop to human events; second, territory is mistaken for normative arguments about the organization of politics; and, third, the territorial state is declining in significance as a geopolitical actor.

The first problem with studies of territory is that they tend to treat areas as objects of human action rather than as an important element in human relations. Indeed, territory provides not just physical or spatial context to human action but also a real embedding of power relations. This idea was termed human territoriality by Robert Sack, who described 'the attempt by an individual or group to affect, influence, or control people, phenomena, and relationships, by delimiting and asserting control over a geographic area' (Sack 1986: 19). Territory, therefore, is a means by which we define relations both among humans and between humans and objects within an area, communicating these relations by marking boundaries and stabilizing them under a regime of enforcement. Sack's work accomplishes two important things by first recognizing that territoriality, understood in this way, is an important feature that sets humans apart from most animal behaviour related to territory since human territoriality is a cognitive function and animal territoriality is instinctual. This serves to redirect studies of aggression and peace towards cognitive structures that are potentially constructive rather than evolutionary arguments for inevitable conflict. Second, Sack provides a useful theory for studying territoriality across different historical, cultural and scalar conditions.

A second problem with contemporary understandings of territory is that their normative assumption of a world of discrete states limits our ability to explain the messiness of world politics. Rather than seeing the world as it is – many multiscalar overlapping and intersecting circuits of power that defy state boundaries – our ideal of state territory as a container limits our focus to a subset of neatly delimited sovereign entities, each with a discrete inside and outside (Walker 1993). This 'territorial trap', as John Agnew terms it, leads to a state-centric vision of politics with three assumed features of the international state system. The first feature is our understanding of power within the state as simply coercive; forcing its subjects to comply with sovereign authority (Agnew 2003c: 55–57). The second feature extends from the first in that we assume relations between states are also based on coercive power, although foreign relations are thought to be different from domestic ones. A third feature of trapping politics in the territorial state is that it tends to ignore the role of state institutions in the creation and extension of capitalist relations both within and beyond state boundaries. Instead, Agnew reminds us of a larger frame for political events where power is also a transformative circulation of resistance and assent between citizens and state institutions both within and between polities, and where capital accumulation and material flows variously rely upon yet defy bounded sovereignty.

A third problem arising from state-centred approaches to territory is that we miss other territorial formations created by non-state political agendas. Some authors have attempted to provide a wider vision and new vocabulary for these non-state territorial politics. Most of these accounts begin with the assumption that the territorial state is in eclipse as the most significant geopolitical actor, or at least that non-state territorialities and non-state actors are of increasing importance. Anxieties about the demise of what one author called the 'hard shell' protecting the nation-state are not new, and partially arise from the realization that economic relations, ideological movements and atomic warfare – not to mention environmental catastrophes – pay little attention to state territories and boundaries (Herz 1957). Timothy Luke has similarly described the deterritorializing influences of contemporary non-state forces ranging from guerilla forces and labour migration to illicit trade and transnational corporations (Luke 1996). Zones of political autonomy lacking statehood and networks of non-state movements across boundaries raise questions about the value of thinking that territories are bounded and singular sites of authority, rather than multiform, overlapping and dynamic sites of political action. These political units are thus reterritorializations not only of non-state and quasi-state entities, but of our notions of states as well.

Case Study: Ethnic Cleansing in Bosnia and Herzegovina

The war in Bosnia and Herzegovina (hereafter Bosnia) provides numerous examples of the importance of territory in understanding contemporary conflict.[2] Although the politics of the war in Bosnia are most readily understood as violent nationalism, this depiction tends to overshadow the territorial vision of what the nationalists hoped to achieve. Nationalists in Slovenia, Croatia and Serbia sought to take their republics out of the Socialist Federal Republic of Yugoslavia, in which they and Bosnia were among six republics. Further, Serb and Croat nationalists wanted to carve up Bosnia's territory (Figure 6.1). However, Bosnia's Croat, Serb and Muslim population had lived together for centuries. The war's hallmark violence against civilians known as ethnic cleansing was to be conducted by militias under the nationalists' operational principle of territorially dividing Bosnia by first forcefully separating its multiethnic population.

Thus, the Serb nationalists in Croatia, Bosnia and Serbia had a goal of creating a greater Serb homeland that required fundamentally reorganizing Yugoslavia's political space. This began when Serb politicians organized local administrative territories into break-away regions,

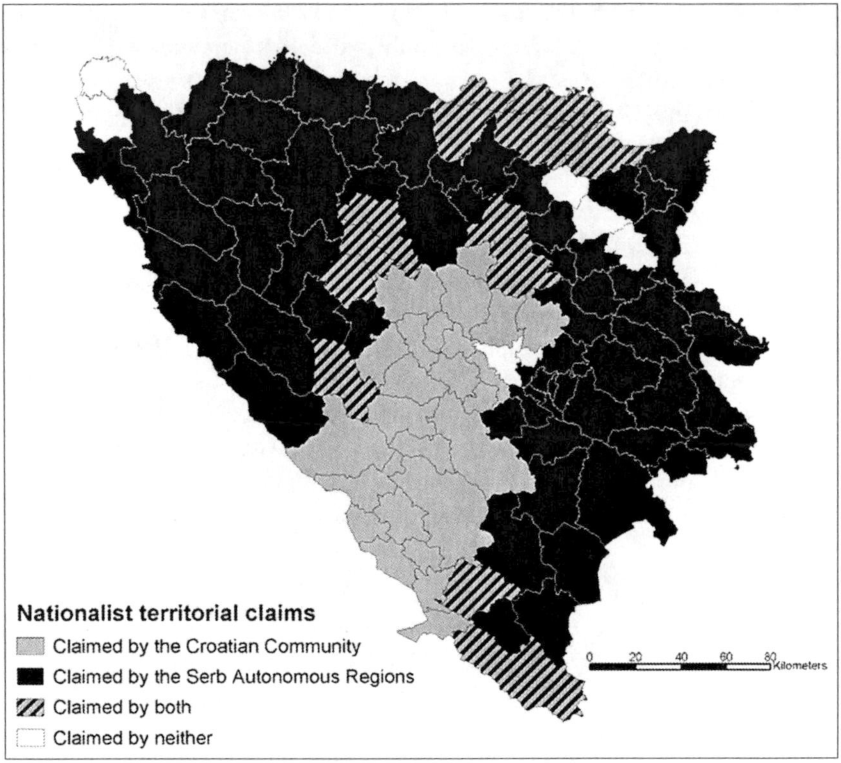

Figure 6.1 Approximate extent of territorial claims made by Bosnian Croat and Bosnian Serb nationalist parties against the territory of Bosnia and Herzegovina on the eve of the war.
(Mahmutćehajić 2000: 132. Cartography by Carl T. Dahlman)

claiming that they would not remain as minorities in a soon to be independent Croatia or Bosnia. These territorial units had been chosen because they created seemingly legitimate local majority claims to self-determination; read another way, this served as political cover for nationalist claims to Serb lands, regardless of the significant number of Muslims, Croats and minority peoples who lived there. Once the wars broke out, first in Croatia in 1991 then in Bosnia in 1992, these breakaway regions formed the basis for independence-minded Serb republics in both countries. These secessionist republics included not only large numbers of non-Serbs but also localities that lacked Serb majorities: in both cases the chosen political solution was to change the ethnic composition of the territory through ethnic cleansing. The Croat community in Bosnia followed a similar path in declaring a separate Croat republic in Bosnia, intent on independence.

Bosnian Serb and Bosnian Croat nationalists saw their independence from Bosnia as the first step in the eventual annexation of their respective territories to Serbia and Croatia proper. For its part, and this is often overlooked, the Bosnian Muslim leadership sought to retain the

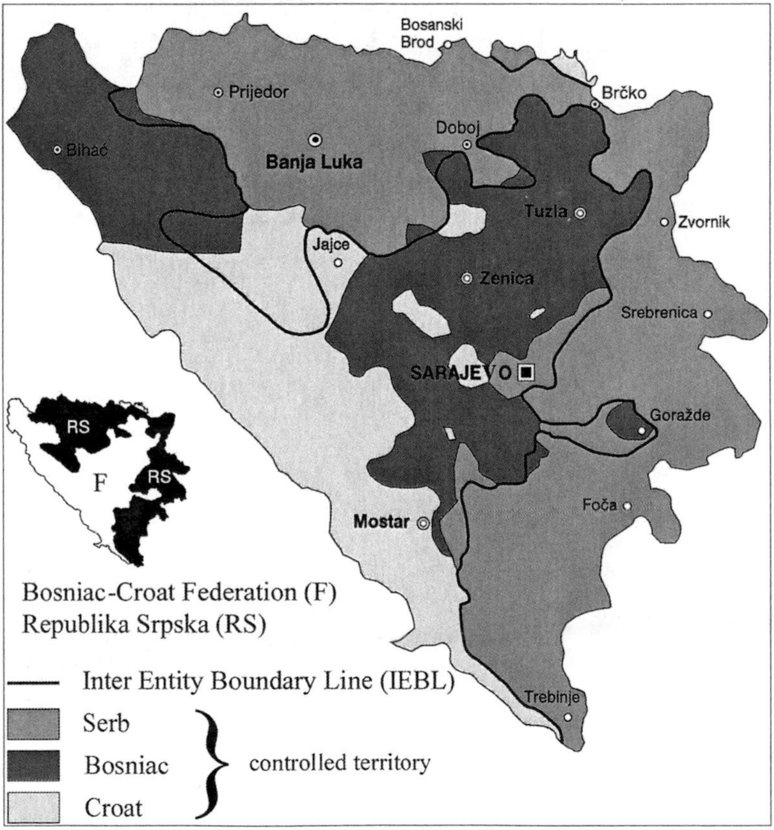

Figure 6.2 Ethnic cleansing forcibly separated Bosnia's multiethnic population during the war. The Dayton Peace Accords produced two territorial entities, one of which was the Serb-held Republika Srpska (Serb Republic) of Bosnia-Herzegovina. The other entity, the Federation of Bosnia-Herzegovina, was internally divided into Croat and Bosniak held areas. (Courtesy of the Office of the High Representative).

Figure 6.3 The Dayton Agreement also established a NATO protected Inter-Entity Boundary Line between the entities, which the Republika Srpska wanted recognized as an international border. This 'welcome' sign remained in place long after it was ordered removed by the international community. (Photo by and courtesy of Carl Dahlman)

territorial integrity of Bosnia against the partitionist strategies pursued by the other parties to the conflict. Moreover, Bosnian Muslims lacked a nationalist tradition and only adopted a distinctive ethnonational identity in the midst of the war, so-called 'Bosniak', predicated on the 1992 boundaries of the Bosnian republic. These competing partitionist versus integrationist approaches toward Bosnian territory were joined by diplomatic interpretations of the conflict. However, the various peace proposals put forward from the outset largely accepted the nationalists' vision for a post-war Bosnia divided by ethnic territories. Territorial assignments in these proposals accepted the premise of the nationalists' claims to separate ethnonational regions, awarded primarily on the basis of which army controlled them.

One such proposal, the Dayton Peace Accords, became the basis for ending the war, triggering a massive international state-building project

Figure 6.4 This billboard posted near the town of Zvornik on the Bosnian-Serbian border encourages residents to call a hotline to identify indicted war criminals. (Photo by and courtesy of Carl Dahlman)

erected atop an ethnonationalist division of territory, population and political authority (Figure 6.2). The treaty sealed the *de facto* partition of Bosnia into two near-sovereign entities, the Bosnian Serb Republic and the Federation of Bosnia and Herzegovina, the latter a tenuous agreement between Bosnian Croat and Muslim parties and internally segregated by ethnicity. The entities were divided by a NATO-patrolled 'zone of separation', a demilitarized area that effectively served as a hard border, severing Bosnia's vital linkages (Figure 6.3). One highly contentious territorial question at the Dayton talks was put off until a post-war arbitral commission could decide what to do with Brčko, a narrow stretch of land that links the two larger portions of the Bosnian Serb Republic, which it conquered by the brutal ethnic cleansing of its local pre-war majority Muslim and Croat population. The solution was again territorial but this time adopted the concept of condominium, shared sovereignty over the same territory, to bind the entities together. Although flawed from the outset and continually challenged by nationalists, Dayton's international peacekeeping and state-building provisions eventually preserved a territorially intact state. Success in promoting the return of those forced from their homes by ethnic cleansing was a slower process, however; it took nearly a decade to achieve even partial success. Yet returns have served to temper, though not defeat, the ethnoterritoriality of the belligerents (Figure 6.4).

KEY POINTS

- Territory is the area claimed or controlled by a state or other political actor. The contested basis for such claims in large part shapes the politics of conflict.
- Contemporary international law provides the basic parameters for state territory, as well as for communal territory such as the high seas or outer space.
- Political geographers also study territoriality as the means by which political actors translate their control over a territory into control over people, resources and relationships.

NOTES

1 The remaining exception to this is Antarctica, which is only partially claimed and those claims are suspended by treaty.

2 This section draws from the author's research. See, for example, Dahlman and Ó Tuathail (2005).

FURTHER READING

Delaney, D. (2005) *Territory: A Short Introduction*. Malden, MA: Blackwell.

Gottmann, J. (1973) *The Significance of Territory*. Charlottesville: University of Virginia Press.

Sack, R. D. (1986) *Human Territoriality: Its Theory and History*. New York: Cambridge University Press.

Taylor, P. J. (1994) 'The state as container: territoriality in the modern world-system', *Progress in Human Geography*, 18(2): 151–162.

7 GEOPOLITICS

Carl T. Dahlman

Definition: Giving Space to Power

The two brief definitions of geopolitics found in the current *Oxford English Dictionary* (2007) tell us a lot about the term's intellectual history. It is, on the one hand, 'the influence of geography on the political character of states' and, on the other, 'a pseudo-science developed in National Socialist [Nazi] Germany'. Both definitions have troubling connotations – the former for its deterministic view of geography, the latter for its connection to fascism.

Geopolitics was originally a Swedish term, coined in 1899 by Rudolph Kjellén, a political scientist who sought to elaborate Friedrich Ratzel's idea that the state was an organism.[1] Geopolitics was thus conceived as the effect of natural geographical factors on the state as a living thing. These ideas proved influential to the thinking of Germany's General Karl Haushofer between the first and second world wars. Haushofer and others were looking for a 'scientific' justification for how Germany could reverse its losses from World War I. Viewing the state as an organism that needed to expand to survive fitted neatly with Nazi plans for territorial expansion, although it is debatable whether these ideas directly influenced Hitler (Bassin 1987; Tunander, 2001). As a result, the term geopolitics became closely associated with the Nazis and fell out of popular use among English-speaking scholars.

The association of the word with the extremes of World War II is too narrow an interpretation, however, since British and American writers were writing about the influence of geography on states at the turn of the last century. Two examples will suffice to show that geopolitics, by whatever name, was a going concern well before the Nazis. Writing during the 1890s, US Admiral Alfred T. Mahan published an influential study in which he argued that a state's geographical location, size and population influenced its ability to become a major sea power (Mahan 1890). Likewise, in 1904, Halford Mackinder, a British academic, sought to explain why the Russian heartland made it a land power

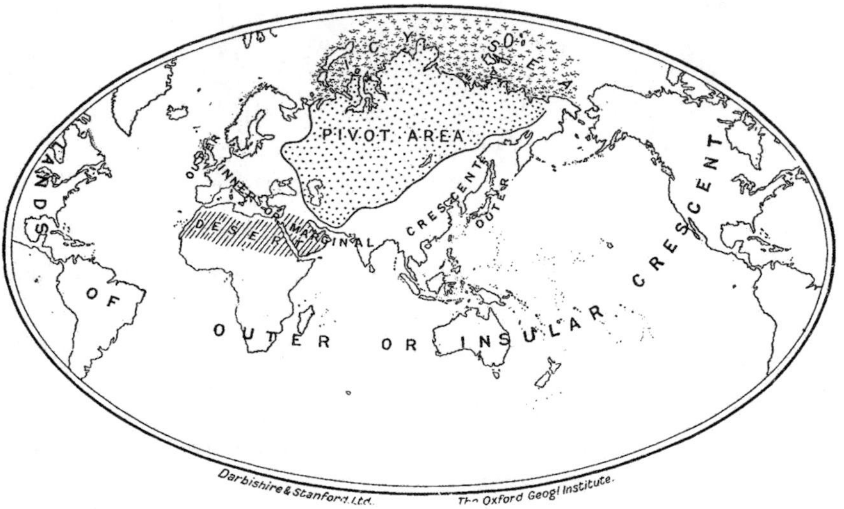

THE NATURAL SEATS OF POWER.

Pivot area—wholly continental. Outer crescent—wholly oceanic. Inner crescent—partly continental, partly oceanic.

Figure 7.1 The Pivot of History was how Mackinder described the vast Eurasian interior (reproduced with permission from Mackinder, H. (1904) The geographical pivot of history, *The Geographical Journal*, 23(4), 435).

that rivalled British sea power (Mackinder 1904). Mackinder argued that as the vast Eurasian interior was traversed with railroads, Russia would be able to push its influence into other regions, from the Black Sea to Japan (Figure 7.1). If Germany gained alliance with Russia, Mackinder worried, they could gain unrivalled global influence at the expense of British interests.

For their part, political geographers after World War II concerned themselves primarily with the *OED*'s first definition of geopolitics as the influence of geography on politics while studiously avoiding the term 'geopolitics', *per se* (Hartshorne 1950). For them, the causal relationship implicit in the definition of geopolitics depends upon the idea that geography is merely the natural condition of the world and that it shapes or determines human action. Indeed, geopolitical writers have often incorrectly cast nature as an active force in human affairs, drawing heavily on the logic of environmental determinism which assumes that cultural, political or racial 'behaviour' is the result of natural causes, such as climate or terrain. In short, geography has been used as an explanatory variable *par excellence*, allowing scholars and policy-makers to neatly tie up loose ends in a variety of political arguments by positing location, borders, and access to the sea or natural resources as the definitive reason for a given political outcome. This crude formula has prevented, and continues to prevent, a detailed understanding of the way geographical factors actually play into political agendas and decision-making.

Even more troubling is that the most problematic aspects of geopolitics are what make it so appealing to so many audiences: an apparent master vision of world politics based on presumably 'hard nosed' geographical facts. As a discourse of mastery, geopolitics is popularly seen as the lofty domain of statesmen, stemming largely from the re-popularization of the term by Henry Kissinger (Ó Tuathail et al. 2006). His writings are typical of the realist school of international relations that paints a world picture of states pursing their self-interest in an anarchic system. States achieve their goals, according to realists, by the application of power, that is, by the coercion or accommodation of other states. The map of world politics that emerges from a realist rendering of international politics is both particular and simplistic. It is particular because it presents the viewpoint of a small number of states, usually just one, ignoring institutional and non-state actors, yet purporting to be an objective bird's eye view of reality. It is simplistic because it reduces the complexity of international politics to vast blocks of space, while presenting itself as a complete rendering of the world. This gives the geopolitical writer an apparent 'prophetic vision' for guiding state policy internationally (Ó Tuathail et al. 2006).

The problems of discourse and presentation, however, are secondary to how geography is 'boiled down' in geopolitical thinking. And this brings us to the fundamental question in geopolitics: exactly how does geography affect politics? Earlier writers like Mahan and Mackinder saw the world from the perspective of imperialism, in which natural advantage gave rise to national advantage. (For more information on imperialism see Chapter 9.) A country's relative position, access to the sea, climate and terrain were indicators of its geopolitical potential, a means of dealing with threats and opportunities. Although the overt use of environmental determinism as a basis for political analysis was common in the early part of the last century, geopolitical writers continued to make arguments based on 'natural fact' to understand the Cold War (Kennan 1947). Communist ideology, for example, was thought to find fertile ground where crops didn't. Contemporary geopolitical visions are no less dependent on apparent geographical causes, as with Jeffrey Sachs's popular take on global poverty as resulting from a disadvantageous location rather than imperial legacy (Sachs 2005). The uneven distribution of oil reserves is frequently touted as a primary if not sole cause of recent conflicts. In a similar fashion, ethnicity has become the new catchall for explaining social conflict based on the assumption that cultural differences, no matter how small, cannot coexist in the same space (Huntington 1996). This horrendous fallacy – that all members of an ethnic group are motivated by hatred – is possible only if the empirical reality of coexistence is ignored (Gagnon 2006).

To be sure, international politics does not play out on a featureless, frictionless surface. The location, resources, population and borders of a state *do* matter but how they matter is as much a function of political interpretation and will than of any determinant fact of nature. So, too, is international politics rendered in terms that are wholly state-centric, failing to recognize the significance of regionalist movements, transnational civil society and the transformative effects of economic globalization (Agnew 2003c: 51–66; for more information on globalization see Chapter 14). Our technologically advanced and interconnected world also means that oil, water and food are globalized commodities that move across borders. Although the geographical facts of nature play an important role in the uneven distribution of these resources, their uneven production and distribution is also a function of a world that humans have made. For example, neocolonial trade practices mean that natural resource endowments provide greater utility to a state in Western Europe than to the community from which they are extracted in the developing world (Slater 2004). Such facts remind us that although the use of the natural world is of human invention, it is no less geographical. Indeed, the geographical structure of power and advantage is the result of a complex relationship between 'facts' and politics. Moreover, these relationships are made known to us largely through our simplified cognitive rendering of the world, returning our attention to the importance of geopolitics as a discursive practice that is, itself, a politics of geographical interpretation (Figure 7.2).

Evolution and Debate: How is Geopolitics (to be) Used?

Two of the most important trends in the field of geopolitics relate directly to the question of how, exactly, geography matters to world affairs. The

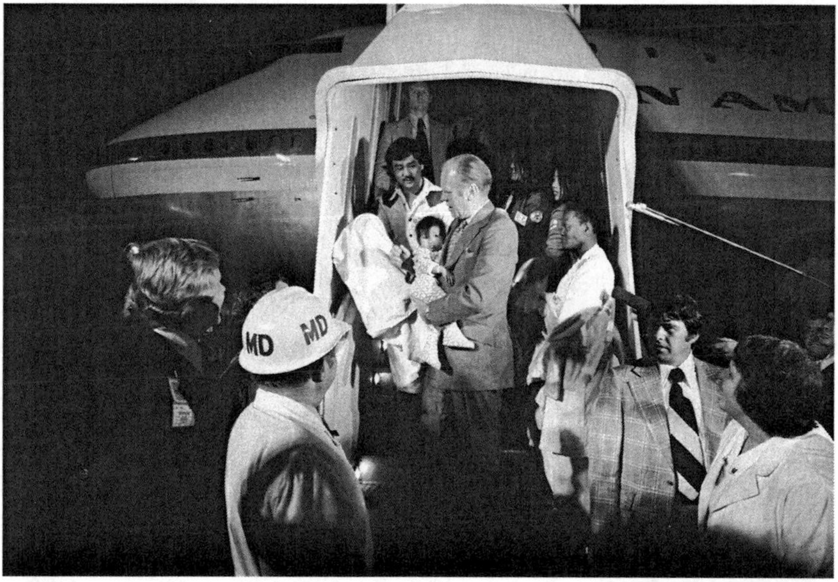

Figure 7.2 The significance of geopolitical interpretation becomes clear in this photo-op during Operation Babylift, in which President Ford is carrying one of several hundred Vietnamese orphans who were evacuated by the US as it withdrew from Saigon in 1975. (Photo courtesy Gerald R. Ford Library)

first of these trends is the emergence of critical perspectives on how geography's apparent matter-of-factness is used in the construction of political projects such as nationalism, security, warfare, globalization and other aspects of international politics. This line of inquiry was advanced by Gearóid Ó Tuathail in his 1996 book, *Critical Geopolitics* that lent its name to this new subfield (Ó Tuathail 1996a). In his approach, geopolitics is not a description of the world as it is but a scripting of the world as statesmen wanted to see it. Geopolitical writings, therefore, are textual mappings of the geopolitician's situated perspective, biases and anxieties. Adopting a deconstructionist approach common to poststructural analysis, Ó Tuathail sought to highlight how these texts construct a world through simplified renderings of messy realities, allowing politicians to act as if those simplifications were reality.

Critical geopolitics soon encompassed a number of authors writing from critical perspectives, though they share no singular methodological or topical interest (e.g. Dalby 2002; Dodds 2000; Sharp 2000). What scholars writing in the vein of critical geopolitics do tend to share, however, is a sense of scholarship as itself a textual intervention, disrupting the easy narratives of state power. They focus instead on the effects of geopolitical statecraft on the lives of those who endure the messy reality it produces.

The second trend in geopolitics asks what is the use and what are the uses of geopolitics? This question motivated a set of essays written in 2002 by four political geographers – Mark Bassin, David Newman, Paul Reuber and John Agnew (Murphy et al. 2004). The question is an important one, considering the intellectual history of geopolitics, extending from its imperial roots, its Nazi connotation and its resurgence as a term used both by leftist academics and realists such as Henry Kissinger and Zbigniew Brzezinski. In his contribution to the forum, Mark Bassin noted that while critical geopolitics had embraced the study of discursive formations, the realists had taken up a 'neo-classical geopolitics', which once again romances a 'geodeterminist logic' built on the 'assumption that geographical space and the natural-geographical world represent objective phenomena which work to constrain and determine the flow of political events' (Murphy et al. 2004: 626). David Newman argues along another dimension to say that scholars of geopolitics, those who take time to look closely at how geopolitical assumptions and decisions differ from reality, have an important role to play in educating a rather poorly informed group of decision-makers, who rely on simplistic and often inaccurate geopolitical constructs. Scholars, in other words, have to make a better 'real world' geopolitics since geopolitical practitioners can too easily make the world worse (Murphy et al. 2004: 626–630). Paul Reuber, however, offers a compelling argument for retaining a decidedly critical geopolitics to counter the fact 'that politics at its core is always geopolitics, since it is a discourse about the spatial division of "us" and "them"' (Murphy et al. 2004: 634). Finally, John Agnew seeks to free the term geopolitics from its close association with geography's troubled disciplinary history of working on behalf of statecraft and to retain it for a broader intellectual use:

> The main reason for wanting to expropriate the word geopolitics is to show how what the word conveys about the dependence of world politics on geographical assumptions and orientations is at the heart of modern practices of world politics, particularly with respect to Great

Power politics. It makes no sense to me to restrict such an evocative word to a relatively minor and uninfluential disciplinary movement (the geopolitics movement) in early twentieth-century geography. (Murphy et al. 2004: 637)

Case Study: Geopolitical Visions of Iraq

Although geography may present a certain degree of a priori factuality, e.g. this is a river, that is a populated place, geopolitics is the act of interpreting those and other 'facts' according to particular and situated interests. We can take as an example the numerous, often competing, geopolitical visions of Iraq produced during the twentieth and early twenty-first century, particularly visions created by British and US foreign policy establishments, including geographers (Figure 7.3). These visions may be subdivided into three sets of interpretations of the Iraqi state. First, there are competing interpretations of Iraq's origins as a state, what one geographer calls the 'artificiality' debate (Stansfield 2007: 26–30). On one side of this debate is a vision of Iraq as an invention of the British following the mandate of 1920 in which the UK subsequently installed the Hashemite monarchy and pushed the country toward independence (Dodge 2003). On the other side is the realization that the mandate was not drawn on a blank map but from existing territorial units that were already somewhat integrated with each other under centuries of Ottoman rule. Moreover, Kurdish and Arab society had extensive trade networks and an extant social order that gave it about as much internal coherence as many of the 'organic' states in the world at the time of their formation (Haj 1997). Nevertheless, a century of self-aggrandizing British historiography and Anglo-American geopolitics has favoured the former interpretation of Iraq as an artificial state. This vision of artificiality has been further exacerbated by an assumption of inevitable ethnic conflict, partially the creation of Saddam Hussein's Ba'th party, which betrays a longer history of more complex inter-ethnic and inter-confessional relations (Baram 1997; Zubaida 2002). Such distinctions fall away in geopolitical discourse, however, as when former Secretary of State Madeleine Albright said a year after the US invasion:

'But we have not shown enough understanding of how Iraq got put together in the first place. I mean, there really are artificial ... it's an artificial country which has to be kept together, but the differences between the Kurds, the Shi as and the Sunnis are massive.'[2]

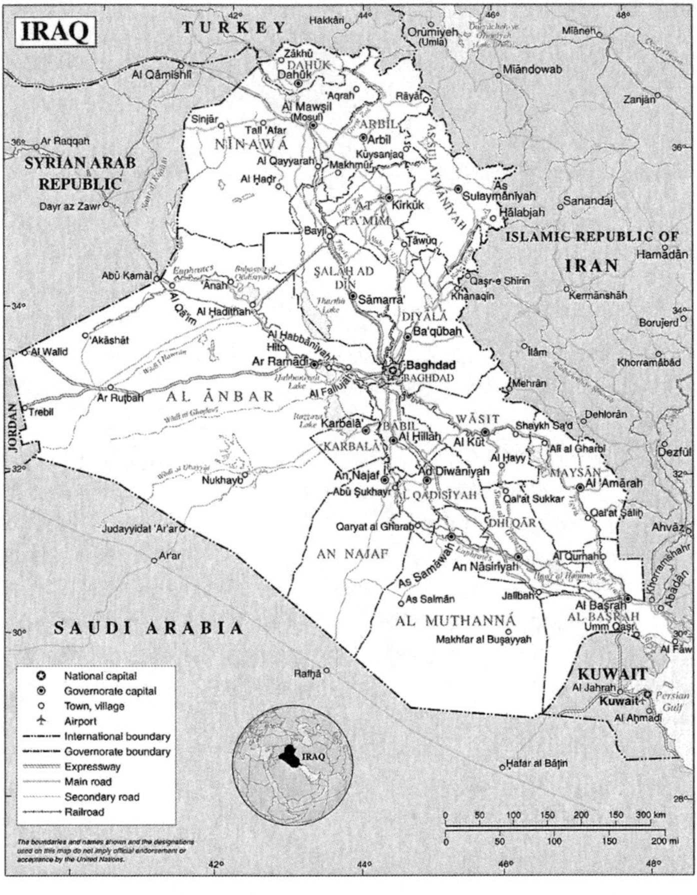

Figure 7.3 Contemporary map of Iraq. Based on UN map 3835, January 2004

British and American foreign policies on Iraq form a second set of geopolitical visions, this time largely realist in nature, that endured through the twentieth century. We can see in the renderings of the Iraqi state by American geographers, for example, just how narrow and superficial are geopolitical facts when it comes to such interpretations. Among the earliest and more influential political geographers in the United States was Isaiah Bowman, who served as an adviser to Wilson at Versailles and later for Roosevelt (Smith 2003). Among his popular writings was *The New World: Problems in Political Geography*, which presented the geography of 'problem areas' that emerged after World War I (Bowman 1921). Bowman viewed Mesopotamia, like Egypt, as a key problem area for imperial Britain, which needed land and sea links to maintain its interests in India and China. Britain's young oil interests at the head of the Persian Gulf were also of concern. Together, wrote Bowman, commerce and oil meant that 'Great Britain's sense of responsibility to the native' was a function of the threat that the 'Mohamadean element' presented to British commercial interests (Bowman 1921: 55, 72). Iraq mattered, according to Bowman, only as it affected British interests elsewhere: 'Mesopotamia, by reason of its position between the industrial nations of the West and the undeveloped populations of the East, becomes a critical region, a true problem area' (Bowman 1921: 74). Bowman glosses over Iraqi society and politics to conclude that should Iraq's Arabs organize or outside powers gain advantage it would injure British, French or Persian interests, thus 'the peace of the world demands that the region be held, if not by England, at least by some other strong power' (Bowman 1921: 74).

Despite their aversion to the term geopolitics, mid-century American geographers like Richard Hartshorne nonetheless tried their hand at explaining the influence of geography on politics (Ó Tuathail 1996a: 158–160). In place of the organic metaphors, Hartshorne adopted ideas drawn from physics such as centripetal and centrifugal forces to describe the forces that bind together or pull apart states. The main centripetal force is the state-idea, a concept that justifies the state's existence and is necessary for Hartshorne because 'the state, to survive, must be able to count upon the loyalty, even to the death, of the population of all its regions' (Hartshorne 1950: 111). Hopelessly lost amid his scientistic analogies, Hartshorne admits the problem of actually finding a state's 'genetic' origins, but he was not shy about explaining Iraq's as:

> (1) the recognition by the Great Powers of the special strategic and economic significance of the Mesopotamian region, and (2) the need to provide a pied à terre for Arab nationalism banished from Syria. (Hartshorne 1950: 112)

Because the idea of Iraq was 'external', created by 'foreign diplomacy and transported nationalist fire', he offers the cautious advice that 'one would need to determine whether the Iraqi [*sic*] have since evolved a truly native concept' (Hartshorne 1950: 112). Although Hartshorne sought to make political geography less 'political' and more objective, he still shared the realists' assumption that Iraq was both artificial and lacking in a sufficiently popular domestic identity. His methodological retreat from politics caused him to overlook an active Iraqi political scene that also downplayed British machinations in the country (Ó Tuathail 1996a: 158–159).

A third set of geopolitical visions has emerged with US military involvement since the 1991 Gulf War and the 2003 Iraq War. One vision, formed as part of the so-called neo-conservative agenda, viewed Iraq as merely one problem in the larger Middle East. In this, the entire region was seen as an artificial creation that could be properly reordered under American guidance to make it safe for allies and commercial interests by reducing threats from Iraq, Iran, the Palestinian Territories, Syria and, for some, Saudi Arabia (Frum and Perle 2004). The agenda is not as coherent as its proponents and critics suggest, but it did recast Iraq in two ways. First, it transformed into an issue of US national security what might be, but was not, going on inside Iraq with regard to weapons programmes, and that served as Bush's justification for pre-emptive war. Second, behind this agenda was the idea that a regime change in Iraq would provide the 'tipping point' for the outbreak of democracy, free trade and stability in the region. This policy, if it was ever so clearly envisioned, has failed miserably. More significantly, the failure of neo-conservative 'idealism' has reinvigorated the enduring 'realist' vision of Iraq, which again ignores empirical facts for the simplicities of seeing Iraq as an ethnically divided, failed state, too important to leave without a strong power for the 'peace of the world'.

A contrasting vision of Iraq, albeit an internally varied one, comes from academic critical perspectives on the US-led war as an exercise of empire (Hardt and Negri 2004; Harvey 2005). This vision brings us back to questions of a far-reaching geopolitical power, namely the United States, and its ability to order the world to its liking (Figure 7.4). Critical geopolitics comes closest to this position not simply in a general opposition to the war but also in returning to the imperial themes of commerce, oil and state-building. The important difference today is that Anglo-American geographers have mostly distanced themselves from the apparatus of their foreign policy establishments. Instead, they maintain an overriding concern for human rights, economic justice and the environment that was not the focus of the state-centred geopolitical views of their predecessors. Derek Gregory, in *The Colonial Present*, rakes over the coals those disastrous British and US policies that have once again made Iraq fail. In so doing, he points to

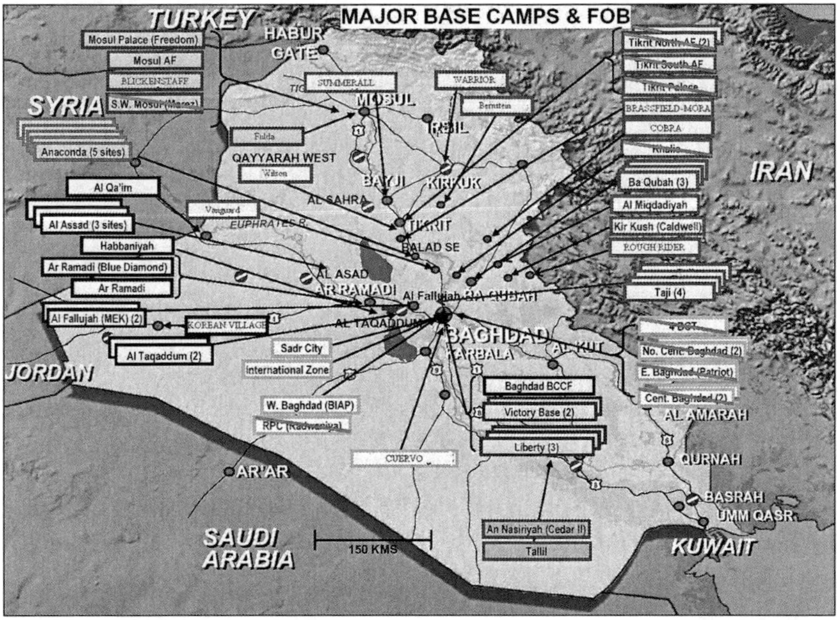

Figure 7.4 US Base Camps and Forward Operating Bases, Iraq, March 2006 (Courtesy US Department of Defense)

the central problem of realist geopolitics, an us versus them dualism that always finds an enemy, even one that's not a threat (Gregory 2004: 262). Exactly how we can escape this spiral of violence is unclear except that the geopolitical apparatus of government–military–capital seems unable and unwilling to change its way of seeing the world.

KEY POINTS

- Geopolitics is the study of the relationship between geographical features and international politics. It is a matter of debate whether these features are natural or constructed, what is meant by international politics and whether the former influences the latter or vice versa.
- Since the late nineteenth century, 'geopolitics' has encompassed several traditions or schools of thought, including one closely associated with the Nazis. This reminds us that geopolitics is also a field of knowledge that developed in the service of state power.

- Recent scholarship takes a critical approach to geopolitics, seeking to unveil the manner in which politicians discursively construct geopolitical spaces, often by manipulating 'geographical facts' for strategic purposes.

NOTES

1 Ratzel (1844–1904) was an influential German geographer.
2 Madeleine Albright interviewed on ABC's *This Week* with George Stephanopoulas, 4 April 2004.

FURTHER READING

Agnew, J. A. (2003) *Geopolitics: Re-visioning World Politics*, 2nd edn. New York: Routledge.

Dodds, K. (2007) *Geopolitics: A Very Short Introduction*. New York: Oxford University Press.

Ó Tuathail, G. (1996a) *Critical Geopolitics: The Politics of Writing Global Space*. Minneapolis: University of Minnesota Press.

Ó Tuathail, G., Dalby, S. and Routledge, P. (eds) (2006) *The Geopolitics Reader*. New York: Routledge.

8 SUPERPOWER

Carl T. Dahlman

Definition: All Bombs and No Brains?

A superpower is a country with an exceptional capacity for successfully pursuing its interests around the world. The earliest use of the term in a geopolitical sense conveys a military superiority that differentiates a superpower from other countries. Namely, a superpower is one that can project its military might and hence diplomatic interests into more than one theatre of conflict, making it a world power rather than a regional power (Fox 1944). The term comes to us more as convenient shorthand for the Allied powers at the end of World War II rather than as a meaningful analytic term. Indeed, until the 1960s, the term 'superpower' was more commonly a description of a spatially extensive, high-voltage electrical network. The analogous application to geopolitics suggests that becoming a geopolitical superpower is a technical phenomenon. In this sense, the term has the quality of natural matter-of-factness, somehow ideologically inert, an inevitable consequence of a country's 'good fortune' in location and resources. The reality is that the first so-called superpowers in the twentieth century – Britain, the United States and the Soviet Union – achieved their position not least because of their geopolitical ambitions and by a coercive application of military force in their self-declared national interest (Figure 8.1).

Considering the more narrow definition of a country that can project its power into multiple 'theatres', it is useful to consider the similarities between a twentieth-century superpower and its historical antecedent, an imperial power. The imperial age of geopolitics is one in which the great powers of Europe competed globally for economic and positional advantage over colonies and markets. In this sense, the imperial powers were each able to pursue their interests in multiple theatres, although they could not sustain a serious challenge by an imperial rival for too long. The exception to this was the British Empire at its zenith,

Figure 8.1 American superpower status was closely tied to its atomic and later nuclear weapons, used here for the second time on the Japanese city of Nagasaki. (Courtesy of National Archives and Records Administration)

which through the use of superior naval technology and economic influence maintained an advantage over other imperial powers. British hegemony during the age of empire thus bears striking similarities to the notion of superpower and these terms remain entwined today. (For more information on the concept of hegemony see Chapter 5.) At the close of World War II, in fact, William Fox first used the word superpower as a geopolitical term to describe a great power with great mobility, embracing Britain, the United States and the Soviet Union.

Britain's retreat from the world stage after the exacting toll of World War II left the United States and the Soviet Union as superpowers in what came to be known as the Cold War. This new bipolar world order was predicated on the two superpowers pursuing their interests in three ways. First, the superpowers divided the world into two exclusive spheres of influence and each sought to prevent the other from adding

Figure 8.2 The 'Big Three," British Prime Minister Winston Churchill, US President Franklin Roosevelt, and Soviet Union leader Joseph Stalin at the Yalta Conference, February 1945. (Courtesy of National Archives and Records Administration)

to its sphere at the expense of their own. This was most clearly articulated at the Yalta Conference where Churchill and Stalin divided Europe into east and west, with the former coming under Soviet control (Figure 8.2). The rest of the world, however, was not so clearly aligned with either superpower. The process of decolonization after the war provided an opportunity for the Soviets and the Americans to enlarge their spheres of influence by providing new states with military, economic and ideological support.

Second, the contest between the superpowers extended into all areas of public life and penetrated deeply into civil society and the private realm. These effects were worldwide but most accentuated among the major countries in each sphere of influence. This was due, primarily, to the often shallow ideological positions encoded into public life: democratic or capitalist freedom among the US and its allies; socialist or communist liberation among the Soviets. Although the superpowers made frequent claims as to the moral correctness of their policies on an ideological basis, the differences were not solely ideological. Decisive in the competition between East and West was the notion that the systems differed in their approach to material and social development. In short, it was a battle over how to achieve a better world, although such ideal horizons were lost in the ugly fight for military supremacy. It was also an expanded form of state competition in which the US and Soviet Union sought to answer the question of how other countries should develop by coercing them into subscribing to certain basic political and economic suppositions. In this way, socio-economic and political development was made subsidiary to geopolitical competition.

Third, the superpowers differed in levels of military ability and position but were equalized by their development of extensive nuclear arsenals. Past a certain point the arms race led to an effective 'balance' of military potential, producing a situation of mutually assured destruction should either the US or the USSR use nuclear weapons. The fear of conventional wars escalating into total annihilation meant that both superpowers and their declared spheres of influence experienced the cold of a war in which no shots were fired between the two sides or their immediate allies. This did not prevent a conventional arms race or military build-ups in Europe, and both countries extended their global reach by investing heavily in naval and airborne technologies. Mostly, however, the superpowers extended their priorities through military alliances that served as the primary means of containment, preventing the expansion of the other superpower's sphere of influence (Gaddis 2005: 20–27). The Cold War was much hotter in the Third World, however, where both powers sought to extend their sphere of influence by means of political sponsorship, economic incorporation, clandestine interventions and war.[1]

It is necessary, moreover, to distinguish the features of hegemony, which is leadership without the necessary resort to force, and superpower, which has the potential to use force without exercising leadership. In practice, the ability of superpowers to forcefully pursue interests with or without allies grants them a *de facto* leadership role, although states may prefer to use 'soft power' and alliances when possible since the price of unilateral action can be quite high. To some extent, geopolitics is concerned with these relations, and the discipline of international relations has spent considerable time trying to quantify and theorize the operation of power in the world. Besides the deeply troubling definitional and operational problems in trying to build models of world politics, these exercises tend to invest in an unproblematic idealization of states as the only relevant actors in an anarchic world system. It might be more intellectually useful to investigate the geopolitical discourse of superpower as well as the particular geographies that have emerged in response to a world with superpowers.

Evolution and Debate: From Geopolitical Fact to Geopolitical Construct

Although superpower is a superlative description applicable to few states, it remains a point of reference for descriptions of 'life at the top' of the geopolitical pile. Its extensive use in popular, practical and formal geopolitical discourse, in fact, underscores its symbolic value in a widely shared imagination of what counts for 'power' in the contemporary world. In other words, the term superpower may not so much tell us about the operation of world politics as lay bare our assumptions about how the political world works. To unpack our implicit expectations through the term it would be useful to draw upon some of John Agnew's work, which deciphers the modern geopolitical imagination (Agnew 2003c). First, 'superpower' implies a global reach, or at least the ability to project influence at multiple points around the world. A superpower thus marks the height of political power within earth's closed space, that visual frame of reference common since the European voyages of discovery and early modern cartography represented the world as a finite set space. That visualization of a world map has also a political hierarchy of places – decided by civilizational, national and technological comparisons – through which a superpower can move as it pursues its interests. This also serves to fix 'lesser' places to regional, national or local areas of influence as appropriate to some measure of their power.

Second, in geopolitical discourse superpowers are defined as the apogee of the 'modern'. They are technically advanced not just in weapons systems but in all fields of science, medicine and technology. The assumption of superpower modernity meant that 'backward' or undeveloped places could be placed in a global hierarchy of geopolitical spaces. For Agnew, this rendering wrongly transformed different spaces into different times: backward places were stuck in earlier stages of development from which great powers and superpowers had emerged long ago. In the ideological programmes of the Cold War, therefore, the apparent lack of civilization of the Third World merely confirmed superpower superiority (Agnew 2003c: 46–47). In short, Third World development was also a geopolitical construct – the superpowers were competing to demonstrate that 'backward' countries could progress best by either a capitalist or socialist political economy. The reality, of course, was that superpower modernity was actually quite limited and rarely transferable in ways that fundamentally improved lives in client states.

Third, superpowers differ from the classical view of a nation-state, sovereign within its boundaries but operating in an anarchic international arena. For Agnew, the conventional mode of state territoriality is problematic in that states are actually highly permeated by transnational economic, cultural and even political forces. Superpowers are in fact a bundle of such forces, penetrating other states' supposedly inviolate affairs either through military coercion, political intervention, economic expropriation or cultural exports. Here again, the difference between superpower and hegemony is blurred except that in the popular imagination superpowers are defined as coercive, whereas hegemony implies broader, more subtle, roles in institution- and norm-building with other states.

Fourth, Agnew's description of the 'pursuit of primacy' applies to the competition between the superpowers for mastery within their respective spheres of influence. Whereas states and conventional great powers are assumed to compete within an anarchic space for power and influence, superpowers have achieved primacy, at least as much as can be expected. Having done so, they then seem to lessen the anarchic qualities of states in their sphere by organizing collective security tasks and economic relations.

Despite their apparent super-abilities, superpowers do not have limitless powers of coercion, as one might infer from the label. The Soviet Union, for example, collapsed sooner and faster than anyone could have expected, leaving it a first-rate nuclear power with a third-rate state. The United States has demonstrated that while its military and communication technology has global reach, its actual grasp remains quite limited, as has been demonstrated in Korea, Vietnam and Iraq. With the exception of a 'popular war' or reinstating the draft, there is little evidence that the United States could prevail in more than one major theatre of war, which has been the policy of US administrations since Kennedy.[2] This is because the US has emphasized technological modernization and lower troop levels, making it better able to fight a conventional enemy army than an insurgency (Figure 8.3). Superpowers must continue to modernize, in the

Figure 8.3 New US airpower provides global reach for 'remote controlled' warfare. The $2 billion B-2 stealth bomber can refuel in mid-air to deliver large payloads of conventional and nuclear munitions against targets protected by defensive systems. (Courtesy of US Air Force)

logic of primacy, to maintain an advantage over competitors, but modernity can just as easily be turned against a superpower, as witnessed on 9/11. Equally troubling for superpowers is that the global reach of modern technology and information flows means that they are highly permeated by interests outside their various territories.

Case Study: Is the US a Superpower, a Hyperpower or a Declining Power?

A useful means of assessing the relevance and meaning of the concept of 'superpower' in a post-Cold War setting is to examine the anxieties expressed about who is or isn't one. Despite the ubiquitous proposition that the United States is the sole remaining superpower, it is important to note considerable debate about the meaning of this role. There is a sense that the term superpower no longer captures the magnitude of US power. At the end of the 1990s, French Foreign Minister Hubert Védrine referred to the United States as a hyperpower, a term that captures the fact that US defence spending accounts for close to half of world military spending. But Védrine also meant a country that is 'dominant or predominant in all categories', especially economically, thereby casting superpowers as merely unrivalled military power (IHT 1999). Amy Chua has since argued that historical hyperpowers – otherwise called empires – grew because of their openness to different cultures but failed when they become intolerant and exclusive (Chua 2007). Ironically, leftist critics of US foreign policy such as Noam Chomsky also play a significant part in upholding the idea of almost limitless American military prowess backed by economic greed and national arrogance by making superpower a convenient term for moral opprobrium (e.g. Achcar 2000).

In contrast, there are authors who argue that the US is no longer a superpower, that its heyday has long since passed and we are now entering a period of multipolarity, leaderlessness or international reorganization (e.g. Johnson 2000). These arguments depict the United States as a country that can no longer win wars and which may have stretched itself too thin in the global 'war on terror'. French analyst Emmanuel Todd argues, *à la* Fukuyama, that the democratization of the world makes US leadership unnecessary while its economic fortunes have made it highly dependent on foreign countries (Todd 2004). These factors, for Todd, are hidden behind the 'theatrical micromilitarism' of US military adventures since the Cold War and by its current over-inflated sense of threat from terrorism (Todd 2004: 144). A similar diagnosis is made by Immanuel Wallerstein (2003: 17–27) who argues that the US is a 'lone superpower that lacks true power, a world leader nobody follows and few respect, and a nation drifting dangerously amidst a global chaos it cannot control'. This path to becoming a powerless superpower began with defeat in Vietnam, popular de-legitimation during the protests of 1968, a loss of purpose after the collapse of the Soviet Union and humiliation on 9/11 by an enemy it can't find. And behind it all, Todd, Wallerstein and others suggest the cause lies in the decline of US economic dominance, especially its oil dependency, debt levels, and loss of technological leadership in some sectors (Lombardi 2005).

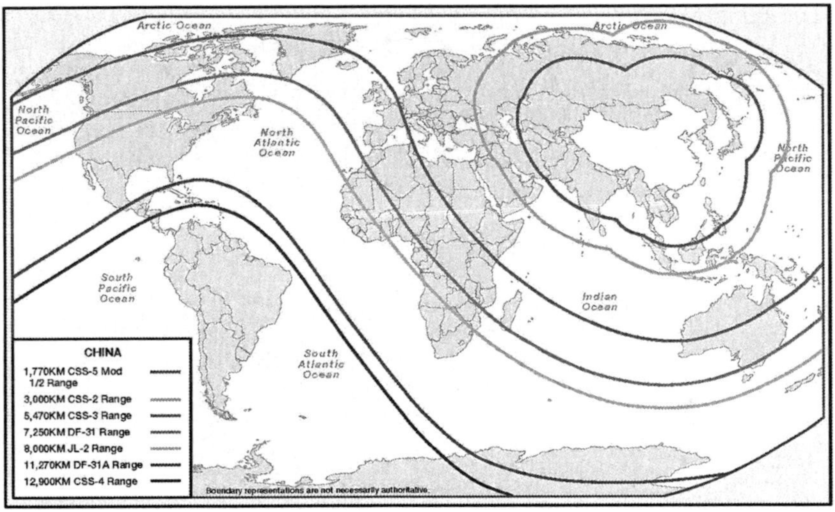

Figure 8.4 Chinese nuclear weapons can now reach targets in most of the world except South America. (Courtesy of Office of the Secretary of Defense)

The foreign policy establishment in the US government is also concerned by trends that might challenge American leadership. The Bush White House implicitly addressed these anxieties in both its 2002 and 2006 National Security Strategies. While remaining preoccupied with the war on terrorism, these documents identify potential challengers to US power, especially China, which is characterized as an emerging 'global player' that must fulfil its 'decision to walk the transformative path of peaceful development' (*National Security Strategy of the United States of America* 2006: 41). The CIA's in-house think tank, meanwhile, depicts trends in China and India as comparable to the rise of Germany and the United States in past centuries. The factors identified as making China and India 'rising powers' are sustained economic growth, growing military capabilities, technological expertise and large populations (Figures 8.4 and 8.5) (NIC 2004: 47–54). It is precisely these trends that some authors seize upon in describing China as a rising superpower, as in the *New York Times* bestseller *China, Inc.: How the Rise of the Next Superpower Challenges America and the World* (Fishman 2006; Liao 2005). There are important reasons to doubt China's interest in extending its sphere of influence beyond East Asia:

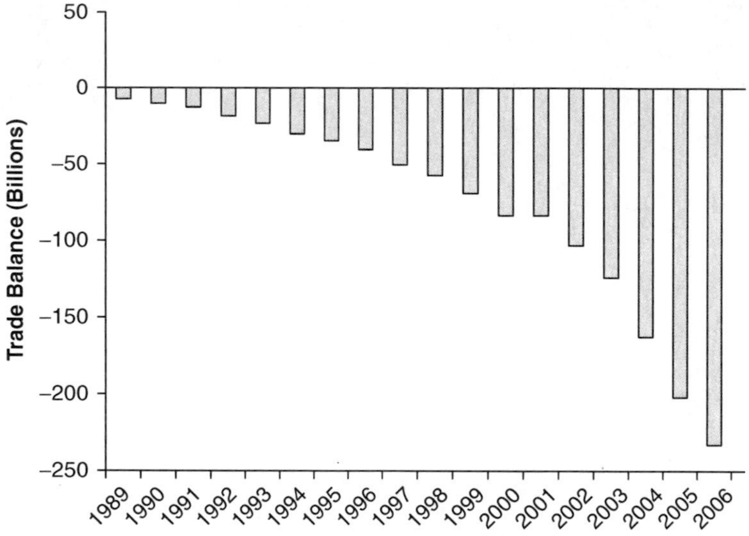

Figure 8.5 A sign of declining U.S. economic power is evident in the rapidly growing trade deficit with China (Table by and courtesy of Carl T. Dahlman. Based on data from US Department of Commerce, 2007).

it is not just that Mao Zedong said it would never become a superpower, but that China continues to face problems with its domestic transformation (Foot 2006). China might do better, some argue, by becoming a non-confrontational and non-hegemonic superpower, meaning it could, in effect, rewrite the script on what it means to be a 'superpower' (Des Forges and Xu 2001).

KEY POINTS

- Superpower is a twentieth century term that describes the unprecedented capacity of a state to pursue its interests globally.
- Superpowers achieve their position through both military and economic superiority.
- Because this concept is built around a small number of cases from recent history – Great Britain, the Soviet Union and the United States – the term is more meaningful as a point of debate than as an analytical category.

NOTES

1 The emergence of the non-aligned movement is critical for understanding this period: see Dodds (2000: 57–60).
2 Binnendijk and Kugler (2001) provide a more optimistic view of US multi-theatre war capacity.

FURTHER READING

Dukes, P. (2001) *The Superpowers: A Short History*. New York: Routledge.
Fishman, T. C. (2006) *China, Inc.: How the Rise of the Next Superpower Challenges America and the World*. New York: Scribner.
Gaddis, J. L. (2005) *The Cold War*. New York: Penguin Press.
Wallerstein, I. (2003) *The Decline of American Power*. New York: The New Press.

Part III
Modernity

INTRODUCTION

Carolyn Gallaher

The modern era is usually viewed as beginning in the late fifteenth century. Most scholars consider Christopher Columbus's trip to the 'new world' as the start of the era. Its end is more difficult to pin down given that some scholars believe the modern era has yet to end. Others, often called postmodernist, believe the modern era ended in the latter half of the twentieth century.

The modern era began with the age of European discovery. The riches these voyages brought back to Europe helped foment the era's most momentous phenomenon – capitalism. This part of the book considers some of the most important concepts related to the modern era.

The first concept covered is **colonialism/imperialism**. The colonial period, which began with Columbus's search for a western route to India, only formally ended in the late 1960s/early 1970s with the decolonization of Africa and Asia. The colonial era was marked by European exploitation in the Americas and later in Africa and parts of Asia. As Chapter 9 demonstrates, the discipline participated in colonial endeavours by providing empires and states with the geographic knowledge necessary for exploitation. In recent years the discipline has started to confront its ugly past, documenting the ways in which it aided colonial-era abuses.

The second concept considered in Part III – **political economy** – discusses the capitalist order that developed in Europe in part as a result of colonial exploitation. The chapter outlines the two dominant views of the then emerging capitalist order, focusing in particular on Adam Smith and Karl Marx. In political geography political economists have been influenced by the largely critical view of capitalism posited by Marx. Political geographers note that capitalism is built on, and relies on, uneven development. That is, capitalist growth does not occur evenly across space but tends to concentrate in certain places.

Chapters 11 and 12 discuss the political thought that emerged in concert with the development of capitalism. The concept of **ideology** captures the ways in which specific social and political interests (normally based on a particular class's interests) are framed, legitimized and normalized. Political geographers have tended to examine ideology as a source of conflict and contestation. **Socialism**, the fourth concept in Part III, is both a particular mode of production/regulation and an ideology.

This part also covers changes in the capitalist order. **Globalization** refers to the compression of time and space afforded not only by technology, but by increasing economic integration. Neoliberalism is a specific approach to economic development in a globalized world. As an approach based on free-market principles (*à la* Adam Smith), this has been widely criticized by geographers for the inequality it has fostered, especially among peasants, minorities and other vulnerable populations.

The final chapter, on **migration**, examines one of the products of globalization (and to a lesser extent neoliberalism). Humans are more mobile than they have been at any other point in history. These flows present challenges and opportunities to those who migrate. Migration streams are also radically remaking the social and economic geographies of the places that migrants leave and move to.

As a group these concepts are very broad. So, too, then, are the case studies provided to illuminate them. They include examples from Mexico, the US and British Columbia, as well as so-called 'spaces in between' where people exist with the aid of technology in multiple worlds all at once.

9 COLONIALISM/ IMPERIALISM

Mary Gilmartin

Definition: Exploitation through Space and Time

The twin processes of colonialism and imperialism have fundamentally shaped the contemporary world. Colonial and imperial relations have influenced political borders, cultural practices and global economic networks. In this regard, European colonialism and imperialism from the fifteenth century onwards has been particularly important. Commentators have identified three broad waves of European colonial and imperial expansion, connected with specific territories. The first targeted the Americas, North and South, as well as the Caribbean. The second focused on Asia, while the third wave extended European control into Africa. Spain and Portugal were primarily responsible for colonizing South America from the fifteenth century onwards, with the territory divided between the two countries by the Treaty of Tordesillas in 1494. They were followed by the British, French and Dutch, who made territorial gains particularly in North America and the Caribbean. Britain – under the auspices of the East India Company – was at the forefront in colonizing Asia, particularly India, from the seventeenth century onwards, though other European countries, such as Portugal, the Netherlands and France, also had Asian colonial possessions. The third wave, from the 1880s onwards, is often described as 'New Imperialism' and was focused on Africa. It was structured by the Berlin Conference (1884–85), which involved the main European powers and served to divide Africa between them. Following the Conference, a number of European powers became involved in the so-called Scramble for Africa. Within twenty years, the vast majority of the territory of Africa had been carved up between Britain, France, Germany, Portugal, Belgium, Italy and Spain (see Thomas et al. 1994). These three broad waves of colonial and imperial expansion have often been linked to the development of capitalism. The first wave has been explained in terms of a crisis of European feudalism, with European powers in search of new sources of revenue. The second wave has been associated with the development of mercantile capitalism and also the development of manufacturing in Europe. The third wave facilitated the consolidation of European capitalism, particularly through the provision of raw materials and new markets (see Thomas et al. 1994; Young 2001). Colonialism and imperialism were highly profitable activities. Among the ways in which these profits were achieved were the appropriation of resources such as silver and gold from the Americas, the sale and use of slaves, trade in spices and opium, and taxes on colonial subjects.

Often, the terms 'colonialism' and 'imperialism' are used interchangeably. Certainly, it is possible to see considerable overlap in their meanings. Both colonialism and imperialism involve the subjugation of one group of people by another, and both processes stretch over space and time, with striking geographical and historical reach. However, a variety of commentators have attempted to distinguish between the two concepts. Lenin suggested that imperialism was the highest form of capitalism, claiming that imperialism developed after colonialism, and was distinguished from colonialism by monopoly capitalism (see Loomba 2005: 10–11). Edward Said wrote that imperialism involved 'the practice, the theory and the attitudes of a dominating metropolitan centre ruling a distant territory', while colonialism refers to the 'implanting of settlements on a distant territory' (Said 1993: 9). More recently, Robert Young suggested that colonialism involved an empire – a group of countries or territories under a single authority – that was developed for settlement or for commercial purposes, while imperialism operated from the centre, was a state policy, and was developed for ideological as well as financial reasons. In this way, Young is suggesting that imperialism is primarily a concept, and colonialism primarily a practice (Young 2001).

Evolution and Debate: A Discipline's Achilles' Heel

The relationship of political geography to colonialism and imperialism is fraught. Whilst some geographers actively participated in the practice of colonialism and the ideology of imperialism, others were active opponents. These competing ideologies have an uneasy relationship, and have gained prominence at different periods in the history of the discipline.

Geographers supported colonialism and imperialism in a variety of ways. Enthusiastic explorers travelled to distant places and returned with tales of their discoveries. These stories were often recounted in the geographical societies of European colonial powers, such as the Royal Geographical Society in London and the Société de Géographie of Paris (Driver 1999; Livingstone 1992). The societies' membership and influence grew as a consequence of their relationship with colonialism and imperialism, and they in turn provided financial support to explorers in search of resources, wealth and knowledge. Moreover, the practice of colonialism was legitimized by geographical theories such as environmental determinism. This theory, which suggested that people's characters were determined by their environment, provided a moral imperative for colonialism. Tropical environments, for example, were seen as producing less civilized, more degenerate people, in need of salvation by western colonial powers. However, the most striking support for colonialism and imperialism came from political geographers, many of whom rose to prominence by providing direct or indirect support for imperial expansion. In Germany, Friedrich Ratzel described the state as an organism that needed to expand in order to survive, in this way justifying imperial expansion. In the UK, Halford Mackinder was an enthusiastic supporter of empire and 'wanted Britain to be the world's greatest imperialist' (Kearns 1993: 17; see Figure 9.1). For many decades, the pro-imperial arguments of Ratzel and Mackinder dominated political geography. Other geographers actively opposed colonialism and imperialism. Other key figures are the anarchist geographers, Elisée Réclus and Peter Kropotkin, who argued against imperialism and nationalism and in favour of less exploitative relations between people and states (see Figure 9.2). The presence of voices critical of imperialism in institutions such as the Royal Geographical Society and among academic geographers has also been noted (Kearns 2004). However, these critiques were often obscured, as political geography celebrated its usefulness to imperial expansion.

Following World War II, the process of decolonization gathered momentum. There were two significant phases of anti-colonial struggle: in Asia immediately after the war, and in Africa from the late 1950s.

Figure 9.1 Halford Mackinder

Figure 9.2 Élisée Reclus

Though anti-colonial struggles led to nominal decolonization, some geographers highlight the ongoing political, economic and cultural legacies of colonialism (see, for example, Berg et al. 2007; Peters 1997, 1998). Geographers and other 'First World' social scientists have also produced critical and theoretically informed studies of European colonialism as a socio-spatial process, often based on fieldwork in post/colonial territories (King 2003). This period saw the advent of the subfield known as tropical geography (which eventually morphed into development geography), which often highlighted and critiqued the ongoing geographies of empire and displayed a commitment to area studies (see Power and Sidaway 2004).

The nature of the relationship between contemporary geography, colonialism and imperialism has come under intense scrutiny. Crucial in this regard is Brian Hudson, whose 1977 article in *Antipode* highlighted the intimate relationship between the 'new geography' and the 'new imperialism' (Hudson 1977). In this article, Hudson argued that academic geography gained importance and recognition because of its usefulness to imperialism, thus directly implicating the discipline of geography in colonial and imperial practices. It took time for his critique to gain prominence in geography, but by the 1990s a range of articles and books had been published that highlighted the myriad ways in which geography and geographers supported and benefited from colonialism and imperialism (e.g. Blaut 1993; Driver 1999; Godlewska and Smith 1994; Livingstone 1992). Similarly, in recent times geographers have written about the spaces of colonialism and imperialism, and the ways in which the material and symbolic appropriation of space was central to imperial expansion and control (see Blunt and Rose 1994; Clayton 2000; Jacobs 1996; Lester 2001). Two excellent examples will be discussed in the next section: Cole Harris's account of colonialism and imperialism in British Columbia, Canada, and Neil Smith's account of the American Empire. Smith's text takes on added significance in the light of a growing number of geographers who argue that we are living under a new imperialism, characterized by accumulation by dispossession (see Harvey 2005 and Chapter 13 on neoliberalism in this volume).

Case Studies: British Columbia and American Empire

Colonialism and Imperialism in British Columbia

Recent work by Cole Harris provides an excellent illustration of the ways in which colonialism and imperialism worked over space and time in British Columbia, Canada. In his book, Harris describes the creation of Indian reserves in British Columbia following the arrival of British colonialists in the province in the 1840s. In this way, the territory of British Columbia was divided into two spaces: one for Native people, the other for almost everyone else. In order to create these spaces,

Figure 9.3 Haida Indian Houses and Carved Posts, Queen Charlotte Islands, British Columbia, 1878. Photo by G.M. Dawson. (Courtesy of Library and Archives Canada)

Native people had to be 'relocated' from their land – often by force – and moved to a number of small reserves created across the province. This was justified by a dominant belief among colonial officials that land occupied by Native people was not being used efficiently and productively (see Figure 9.3). This process was not unique to British Columbia. It also gained expression in the eighteenth-century clearances of the western Highlands in Scotland and in the creation of small reserves for Aboriginals in Australia. Through the creation of reserves, as Harris comments, 'one human geography was being superseded by another, both on the ground and in the imagination' (Harris 2002: xvii). Harris describes the resulting reserve geography as serving two key functions. The first was to open up land in the province for exploitation by settlers and by capital. The second was to replace use rights – where members of a group had access to and use of resources without owning them – with a system of private property (Harris 2002: 266).

By the 1930s, there were more than 1,500 reserves scattered across British Columbia (Harris 2002: 265), most of which were considered too small to provide a viable livelihood for the Native people who lived on them. This was not unintentional, since there was a general belief that smaller, unviable reserves would force Natives into the job market, where they could become civilized. In addition, smaller reserves would pose less of a military threat and would be a source of cheap seasonal labour. The effects of reserve geographies on Native livelihoods were, as Harris points out, far-reaching. The agricultural and extractive possibilities of reserves were limited, so Native people who remained on reserves became incorporated into the waged economy in a very tenuous way. Others became seasonal migrants, while still others moved to urban areas. Conditions on the reserves 'took their toll on Native bodies' (Harris 2002: 291), with high levels of disease and mortality.

Harris highlights the ways in which the history of British Columbia is composed of two competing narratives: development and dispossession. The narrative of development – of a British colonial transformation of wilderness into useful land – dominated. In doing so, it discounted the cost of development for the Native inhabitants of the province. The narrative of dispossession has been predominantly recounted by Native people; it tells of how their land was taken away, leaving them with insufficient land to live on (Harris 2002: 294). Harris is at pains to point out the contested nature of the reserve policy: some colonialists opposed the policy or sought a different, more humane articulation of the relationship between the colonizers and colonized; many Natives fought against their marginalization and argued for their place in the territory of British Columbia. Unfortunately, as Harris notes, these alternative visions were sidelined in the interests of state, capital and racial hierarchies.

American Empire and the Role of Isaiah Bowman

The choice of a text that highlights American imperialism, given the focus on European colonialism in the earlier part of this chapter, needs explanation. Smith's discussion of the role of geographer Isaiah Bowman in the articulation of a new form of empire shows the ways in which the relationship between geography, colonialism and imperialism was continually being recreated and refashioned. Smith describes Bowman's early voyages of exploration and conquest in South America in 1907, 1911 and 1913 as attempts to 'make the world' (Smith 2003: 78). When Bowman became director of the American Geographical Society in 1914, he continued this making of the world through support for exploration and cartographic advancement. He was appointed to President Woodrow Wilson's inquiry in 1917: the role of the inquiry was to gather data that would underpin the US authorship of a 'new world', characterized by geographical order (Smith 2003: 121–122). This new world order was articulated by Wilson and the American delegation at the Paris Peace Conference, and his involvement in the process earned Bowman the title of Wilson's geographer. Later, Bowman was also described as Roosevelt's geographer, when he became a special adviser to Roosevelt in 1938. In the same period, Bowman strategically used American military imperatives to reassert the place of geography in American universities.

Smith's account of Bowman's work has resonance for a discussion of colonialism and imperialism. The question of the status of colonies was central to US negotiations with Britain during the Second World War, and Bowman was a key figure in these. During this period, Roosevelt was vehemently opposed to European colonialism and Bowman was much more circumspect, but both recognized the need to incorporate European colonies into a US-led global economic order. As Smith comments, 'the United States' antipathy to colonialism in this period expressed at root a self-interested drive to open new markets' (Smith 2003: 373). More broadly, Smith argues that Bowman was a central figure in the replacement of a territorially based, European-centred colonialism with a non-territorial, market-led, US-dominated imperialism. Smith describes this as the American Empire.

Both Harris and Smith are at pains to point out the ways in which colonialism and imperialism were always contested enterprises. Their attention to contestation has relevance, not just for understanding how colonialism and imperialism worked in the past, but also as to how new forms of imperialism take root in the contemporary world.

KEY POINTS

- The contemporary world has been fundamentally shaped by the processes of colonialism and imperialism. Though there are differences between the processes, they share some common features.
- Some geographers – such as Halford Mackinder – provided practical and political support for colonialism and imperialism. Others – such as Elisée Réclus and Pyotr Kropotkin – argued against colonialism and imperialism.

- In recent years, geographers have paid particular attention to the ways in which the material and symbolic appropriation of space was central to imperial expansion and control, as they re-examine the relationship between geography, colonialism and imperialism.

FURTHER READING

Blaut, J. (1993) *The Colonizer's Model of the World: Geographical Diffusionism and Eurocentric History*. New York and London: Guilford Press.

Godlewska, A. and Smith, N. (eds) (1994) *Geography and Empire*. Oxford and Malden, MA: Blackwell.

Harris, C. (2002) *Making Native Space*. Vancouver: University of British Columbia Press.

Hudson, B. (1977) 'The new geography and the new imperialism: 1870–1918', *Antipode*, 9(2): 12–19.

10 POLITICAL ECONOMY

Carolyn Gallaher

Definition: The Capitalist Order

Political economy is the study of the principles that govern the production and distribution of goods under capitalist systems. The term originated in the eighteenth century to describe Europe's budding capitalist economies and state efforts to manipulate them for crown and country. As the term suggests, political economy is concerned with more than straightforward economic functions, like supply and demand. It includes the roles that individuals and political entities such as states play in the production and consumption of goods. Its wider focus means that political economy is as common in disciplines like geography, anthropology and sociology as it is in economics.

The two main analytic approaches to political economy were laid out by contemporaries Adam Smith and Karl Marx (see Figures 10.1

Figure 10.1 Drawing of Adam Smith

Figure 10.2 Karl Marx 1866

and 10.2) respectively. Unlike the prevailing wisdom of the day, which held that the wealth of a nation-state was measured by its agricultural development, Smith and Marx located national wealth in the industrial sector. In particular, both men subscribed to the labour theory of value which holds that wealth is produced from the difference between the real value of something – the labour required to produce or get something – and its exchange value – how much a producer could sell a product for beyond its intrinsic value (Dooley 2005). However, both men viewed the overall character of the emergent industrial order, and the rules that governed its production, quite differently. In the review below I only examine the analytic precepts developed by Smith and Marx, though it is important to note that both thinkers also developed related political prescriptions. See especially the *Communist Manifesto* by Karl Marx and Friedrich Engels (1848).

Adam Smith

For his part, Smith believed that the capitalist order was an engine of wealth because it relied on a division of labour. As an example, he proffered the trade of pin-making. Smith noted that making a pin required 18 distinct tasks, including drawing out wire, straightening it, cutting it, pointing it, grinding it, and so on. Smith postulated that a man, working by himself, could make no more than 20 pins a day. When pin production was divided between 10 men, they could produce 48,000 pins a day (Smith 1957).

Smith believed that the capitalist system would produce great gains for society as a whole because people acted in their self-interest. As he notes in an oft-quoted passage from *The Wealth of Nations*, 'it is not from the benevolence of the butcher, the brewer, or the baker, that we expect our dinner, but from their regard to their own interest' (1957: 11). That is, while the butcher may cut and sell meat because he makes money doing so, the net result is that more meat is made available for others to eat.

Smith also believed that the market functions best when it is allowed to correct itself. If the production of a given product trailed demand, a producer would emerge to fill that demand. Likewise, if the production of a product far exceeded demand, its price would drop and its producer would shift production to something with a higher demand and greater price. The market's capacity for self-regulation is often referred to as the invisible hand of the market, although Smith himself only used the term once in *The Wealth of Nations*, and only obliquely in reference to a self-regulating market.

Karl Marx

While Smith's overall assessment of capitalism was generally upbeat, Marx viewed the emergent system with suspicion. According to Marx, capitalist relations represented a first in human history. Whereas individuals in previous economic systems used labour to produce goods for exchange, in the capitalist system labour itself became an object of exchange. That is, people sold their labour for a wage. Marx argued that the commodification of labour led to the alienation of men from the goods they produced (Marx 1961). It also led to the formation of a new class structure based on two groups – the proletariat, who sold their labour power for a wage, and the bourgeoisie, who purchased it and extracted surplus value from it. Marx saw this system as inherently exploitative of labour.

Unlike Smith, who saw the capitalist marketplace as self-regulating, Marx viewed the capitalist system as inherently unstable. He believed the system was crisis-prone because it was beset by internal contradictions (Marx 1962). Primary among them was the impetus to expand the profit margin. Because labour was the source of profit, the bourgeoisie would search for new ways to extract more surplus value from it. Eventually, however, there would be a declining rate of return from such measures and over time the overall rate of profit would decline. Eventually, the rate of profit would reach zero and the system, or portions of it, would collapse. Labour would then become less valuable, and new entrepreneurs would take advantage of cheap labour to invest in new technology and forms of production. Marx argued, however, that over time capitalist crises would grow progressively more severe and eventually workers would unite around their common class interests and overthrow the system.

Evolution and Debate: From the Material to the Symbolic

In geography, political economy approaches have tended to attract scholars from critical wings of the discipline. The discipline's earliest political economists were influenced by Marxism, French existentialism and humanism (Peet and Thrift 1989). Today political economists are also influenced by so-called 'post' forms of analysis, including poststructuralism, postmodernism, postmarxism and psychotherapy (Gibson-Graham 1996; Gibson-Graham et al. 2000).

As a Critique of the Discipline

Political economy first gained prominence in the discipline during the late 1960s and 1970s. Its rise can be understood as both a product of its time – a reflection of the turbulent 1960s – and as a response to almost a century of decline in geographic thought (Peet and Thrift 1989). Some of the earliest geographers to employ political economy did so in the context of political activism. William Bunge, for example, employed political economic critiques to aid activism in inner city Detroit (1969). Other geographers used political economic approaches to understand geography's history and its decline during the first half of the twentieth century (Livingstone 1992; Peet 1985). In a now seminal article in the *Annals of the Association of American Geographers*, Richard Peet (1985) argued that environmental determinism was developed by geographers to further the imperial interests of European and American governments. By labeling people in colonial areas of conquest as morally inferior, geographers helped legitimize exploitation. Environmental determinism served state interests handsomely for many years, but when its intellectual underpinnings were exposed as Eurocentric and unscientific, the discipline's clout as a social science suffered serious damage.

Uneven Development

Geographers have also contributed to the wider political economy approach by expanding its analytic categories. In particular, they have pushed political economists to examine how political economies are manifested spatially (Harvey 1973, 1982; Smith 1984; Wallerstein 1974). Most of these works were Marxian in approach and tended to assert that uneven development was 'a structural imperative of capitalism' (Peet and Thrift 1989: 14). One of the first studies to do so was by Immanuel Wallerstein, a political scientist with geographic sensibilities. Wallerstein (1974) argued that colonial endeavours in the sixteenth century created a 'world system' in which so-called 'core' countries and regions extracted resources and labour from 'peripheral' places. In short, the bourgeoisie/proletariat class structure was manifested not only socially, but spatially.

Wallerstein's analysis was considered radical at the time. It was a pointed critique of scholars, such as Walter Rostow (1960), who argued that the differing pace of development could be explained by the fact that some countries possessed 'natural' advantages over others.

While Wallerstein's notion of core/periphery relations continues to have an indelible imprint on the discipline,[1] David Harvey, a Marxist geographer, is probably the discipline's most recognizable political economist and its most theoretically influential one. Harvey's primary contribution to political economy is his notion of the spatial fix, which he first outlined in his 1982 book *The Limits to Capital* (Schoenberger 2004).[2] He used the concept in two ways.

In the first, Harvey used the term to describe how capitalism dealt with the tendency towards overaccumulation. Marx theorized that as capitalist economies produce more and more wealth, the rate of profit begins to fall. In response, business expands production to keep profits growing. Eventually, however, overproduction occurs, and prices drop to account for the oversupply of goods versus the demand for them. Overaccumulation, as this situation is called, leads to recession or other crises in the system. Harvey (1982) argued that capitalists deal with the tendency towards overaccumulation by employing a spatial fix. In brief, when profits start to decline, capitalists 'extend [the] frontiers' of the region in which they operate or 'move their capital to greener pastures' (1982: 417). The result is uneven development: as some places 'boom' others 'bust', and those that 'bust' can 'boom' again.

Harvey (1982) also uses the term 'the spatial fix' to talk about geographical inertia that develops in capitalist systems. That is, even as capital is footloose and fancy-free, it also requires investment in a stationary built environment, including factories, warehouses, headquarters and the like. States contribute to the fixed nature of capital by building schools, highways and hospitals in the areas where capital and labour concentrate. When the capitalists want to use the spatial fix, however, they often find themselves thwarted by vested interests, including governments, investors and labour. Harvey's dual usage of the spatial fix captures what he sees as an inherent contradiction within capitalism: that capitalism's imperative for mobility often collides with its territorial logic.

Geographers have also applied political economy approaches to other thematic areas in the discipline. Neil Smith's work on gentrification (1996) is a case in point. Smith sees gentrification, a process where older, centrally located urban areas are refurbished by young professionals, as a product of late capitalism. Using examples from the US and Europe he demonstrates how gentrification is a key battlefield in the contemporary class struggle. He also uncovers the discursive alignments that give meaning to the gentrification process. In the US, especially, gentrification is depicted as a civilizing force on the urban frontier – a contemporary Wild West where crime and social disorder are rampant. Gentrification and the discourses that support it feed what Smith calls the revanchist city – a place where urban professionals go to greater and greater extremes to separate themselves spatially and emotionally from the urban 'rabble' (Smith 1996).

Geographers have also applied political economy approaches to studies of human/nature interactions. This approach is often called political ecology because it examines how political economies affect ecological functions (Robbins 2004). Political ecology approaches in geography have examined land degradation (Blaikie and Brookfield 1987) and deforestation (Rangan 2001) among other ecological problems. In 2001 Nancy Peluso and Michael Watts published *Violent Environments* in which they theorized the connection between conflict and the political economies that shape access to resources. They argue against Malthusian approaches which see violence as the result of scarcity. Instead, they demonstrate that 'abundance [of resources] and processes of environmental rehabilitation ... are most often associated with violence' (2001: 6). Political economic systems that exclude social actors from available, and necessary, resources, can and do create conflict.

Unpacking the Economy

In the last decade, geographers have begun to use so-called 'post' forms of analysis to examine political economy. J.K. Gibson-Graham[3] were two of the first authors to use such an approach. In *The End of Capitalism (As We Knew It)* Gibson-Graham (1996) employ a poststructural approach to political economy. Like other 'post' forms of analysis, poststructuralism focuses on the manner in which categories of meaning are constructed, normalized and deployed in language and discourse. In their book Gibson-Graham deconstruct the category of capitalism, arguing it is a monolithic category that obscures what are discrete and often disconnected processes. There is no one thing that can accurately be called capitalism, they argue. Rather, the term captures so many disparate things – different types of economic transaction, labour formations, class concerns and modes of production – as to be all but meaningless. While Gibson-Graham's formulation is unorthodox for traditional political economy, their critique is situated in a classic concern of political economy: to end the exploitation of labour and environment made possible by unequal modes of production. They argue that their desire to break apart the unity of the category is political. Capitalism needs to be unpacked and disaggregated in order to be resisted. As currently constructed, the category of capitalism is too all-encompassing to resist.

More recently, scholars have employed 'post' forms of analysis in other traditional categories of political economy. Studies by geographers have unpacked the category of class (Arvidson 2000), blue and white collar employment (Southern 2000), and slavery (van der Veen 2000) among others. While such studies are relatively new, they will probably continue, given the rising profile of poststructural approaches in geography, specifically, and the academy more broadly.

Case Study: How Do We Explain the Development Divide?

In 1997 evolutionary biologist Jared Diamond published *Guns, Germs, and Steel.* The book, a Pulitzer Prize winner, purports to explain the geographic basis of the development divide. Although the book won plaudits in the mainstream press, geography gave the book a decidedly cooler reception (Blaut 2000; Jarosz 2003).

The discipline's ambivalence towards Diamond's book can only be understood in context. As the work detailed above makes clear, geography's complicity in the colonial project, particularly its attempts to provide a scientific rationale for racial exploitation, almost ruined the discipline, especially in America. By the 1920s scientific analysis was beginning to uncover just how superficial and unscientific theories of environmental determinism were. Geography departments found themselves on the chopping block, and some were eliminated in Ivy League universities and beyond.

The appearance of Jared Diamond's book, then, sent shivers through the discipline. Indeed, the book's driving question – 'Why did history unfold differently on different continents?' – is the very question the discipline's early environmental determinists sought to answer. Diamond cautions readers, however, that the book would not delve into the standard racist fare of the genre: 'In case this question immediately makes you shudder at the thought that you are about to read a racist treatise, you aren't' (1997: 9).

Instead of racial differences, or other 'standard' explanations, like culture, Diamond argues that places that eventually developed had three characteristics in common. First, they had naturally occurring pulses, such as wheat, barley and/or peas. Diamond notes that the seeds in pulses were easier to domesticate than root crops or other staples like maize. Second, places needed megafauna capable of domestication. Europe's wild horses, for example, were capable of domestication while Africa's zebra were/are not. Domesticated animals were vital for settled agriculture because they could be used for ploughing, which substantially increases the amount of land under cultivation. Third, continents with an east–west orientation were at a distinct advantage because this allowed the diffusion of food production technologies. Diffusion on continents with a north–south orientation was impeded by climatic barriers since most crops are not suited to multiple climates. Taken together, the factors allowed permanent human settlements (rather than seasonal ones) to form and afforded, through the crowding of humans and animals, a resistance to germs and the development of technologies such as metallurgy.

Diamond's most trenchant critic in geography has been the political economist James Blaut. Given environmental determinism's reputation, Blaut asks why the theory is making a comeback, with the publication not only of Diamond's book, but of similar treatises, including David Landes' *The Wealth and Poverty of Nations* and Eric L. Jones' *The European Miracle*. Blaut hypothesizes that scholars have re-engineered environmental determinism to focus so heavily on its geographic elements because the former foundations of the theory – European racial superiority and, later, cultural superiority – have proven difficult to sustain. The ultimate goal, however, is to reaffirm western exceptionalism.

Blaut offers a detailed critique of Diamond's book. Only three of his points are reviewed here. One of Blaut's criticisms of Diamond is his tendency to conflate regions in ways that benefit his thesis of European dominance. While Diamond correctly identifies the Fertile Crescent as the first site of plant and animal domestication, for example, he does not explain why it was Europe, rather than the Middle East, that subsequently 'took off'. Indeed, Diamond glosses over the distinction, conflating the Middle East with Europe when discussing the start of civilization, even though he later makes a distinction between the two regions. Moreover, he fails to account for what all of this means for his wider theory. In particular, one might ask whether Europe was actually geographically superior to other parts of the globe or simply better at adopting innovations developed elsewhere.

Blaut also takes on Diamond's assertion that an east–west orientation is vital for technological diffusion. Blaut asks, for example, why climate is considered a barrier to diffusion while mountains, deserts and other physical features are not. Diamond fails, for example, to explain why physical barriers such as the Gobi Desert and the Himalayan and Caucus mountain ranges did not limit the diffusion between China and Europe. In short, Diamond fails to explain why some barriers behave as expected (i.e. limiting diffusion) while others do not.

Blaut also argues that Diamond fails to do what he proposed at the start of his book, which is to 'identify the broadest patterns of history'. In particular, Blaut suggests that Diamond fails to explain why it was Europe, rather than Eurasia as a whole, that became dominant. Indeed, given the emphasis on the importance of an east–west orientation to development, China should have developed as quickly as Europe. Diamond answers this question by arguing that China's lack of barriers facilitated the creation of an empire, which led to despotism and the stifling of entrepreneurial activities. Europe, by contrast, never developed an empire of the same size – its hearth areas remained relatively isolated – and was thus eventually able to morph into a state system. Blaut takes issue with this explanation on a number of fronts. For starters, he suggests that Diamond's geography is all wrong. The physical barriers within contemporary China are not substantially different from those in Europe. Blaut also takes issue with what he sees as a differential view of barriers in Diamond's hypothesis. In China, a lack of barriers is posited as an explanation for its development lag whereas elsewhere in the world their absence is regarded as facilitating development. Diamond fails to explain this glaring contradiction, and as a result the causal effect that Eurasia's east–west orientation is meant to have is called seriously into question.

For Blaut political economy provides the clearest answer to why Europe developed faster than other parts of the globe. Europe was, he

Figure 10.3 The dungeon at Elmina Castle, Ghana housed slaves bound for Brazil and the Caribbean. Portuguese slave traders profited handsomely from the slave trade. The Portuguese state also profited from the labour of slaves once they landed in Brazil. Slaves were used to grow tropical crops which Portuguese traders sold across Europe. The proceeds of these markets were invested in Europe rather than Brazil.

Figure 10.4 In this painting Robert Clive, the British governor of India is seen receiving a decree from Shah Alam, the Mughal Emperor of India, guaranteeing the East India Company the right to administer the revenues of Bengal, Behar and Orissa. Britain developed by extracting surplus from its far flung colonies and investing it at home.

argued, catapulted ahead of its global peers by a simple fact: 'Europe acquired incalculable riches from the Americas after 1492' (2000: 2). These riches, which were often extracted by the forced labour of indigenous people, were invested in Europe, giving rise to a merchant capitalist class and the territorial power that came with it (see Figures 10.3 and 10.4). Unfortunately, Blaut and Diamond were never able to continue their debate. Shortly after he published his critique, Professor Blaut passed away.

KEY POINTS

- Political economy is the study of the principles that govern the production and distribution of goods under capitalist systems.

- There are two broad schools of political economic thought – those that follow Adam Smith and those that follow Karl Marx.
- Political geographers use the concept of political economy in three ways. Some use it to unpack the historical role the discipline played in the colonial era. Others use political economy to understand uneven development. A third approach critiques traditional views of political economy by arguing that capitalism is not a monolithic phenomenon.
- Political economists in political geography do not believe that the development divide between the global north and global south is the result of natural advantages (in terrain, climate, etc.) found in Europe. Rather, they argue that Europe developed before Asia, Latin America and Africa because it stole their resources and invested them internally.

NOTES

1 For a critique of world systems theory see Abu-Loghod (1989). Abu-Lughod argues that there was a world system in place before the period of European colonialism.

2 In later work Harvey uses a broader term – the spatio-temporal fix. While he does not explain the shifting terminology, it may be a response to critics who suggested that he did not give sufficient weight to the temporal aspects of capitalist uneven development. See especially Jessop (2002).

3 J.K. Gibson-Graham is the pen named used by Julie Graham and Katherine Gibson for their collaborative work.

FURTHER READING

Marx, K. (1961) *Economic and Philosophic Manuscripts of 1844*. Moscow: Foreign Languages Publishing House.

Marx, K. and Engels, F. (1848) *The Communist Manifesto*. Retrieved 13 January 2008, from http://www.marxists.org/archive/marx/works/1848/communist-manisfesto/ index.htm

Smith, A. (1957) *Selections from 'The Wealth of Nations'*, ed. by George Stigler. Arlington Heights, IL: AHM.

Smith, N. (1984) *Uneven Development: Nature, Capital and the Production of Space*. Oxford: Basil Blackwell.

11 IDEOLOGY

Mary Gilmartin

Definition: From Idea to Proscription

When the term 'ideology' first appeared in the English language, at the end of the eighteenth century, it was used to describe the science of ideas (see Williams 1988). Over time, ideology came to be understood in a broader sense as an abstract or impractical theory. This is the way the term was used by Marx and Engels in *The German Ideology*, where they suggested that ideology meant illusion, false consciousness or upside-down reality. However, Marx and Engels also suggested another meaning for ideology, that of a set of ideas that arise from a definite class or group (see Williams 1988). In his discussion of the meaning of ideology, Williams suggests that the term is generally described pejoratively, often in opposition to science.

Williams' definition of ideology provides one route to understanding the term. A second approach is suggested by Anthony Giddens, who defines ideology as 'shared ideas or beliefs which serve to justify the interests of dominant groups' (Giddens 2006: 1020). In this way, ideology is linked to hegemony, a concept illuminated by Antonio Gramsci to highlight the way power is applied through consent and coercion. (See Chapter 5 for more detail.)

In his book on the topic, Terry Eagleton identified 16 different general uses of the term ideology, and distilled these into six broad and related definitions. Thus, according to Eagleton, ideology can mean:

- the material process of the production of ideas, beliefs and values in social life;
- ideas and beliefs that symbolize the life-experiences of a socially significant group or class;
- the promotion and legitimation of the interests of a socially significant group or class;
- the promotion and legitimation of a dominant social power;
- ideas or beliefs that legitimize the interests of a ruling group by distortion;

- false or deceptive beliefs that arise not from the ruling group, but from the structure of society as a whole (Eagleton 1991: 28–30).

Within political geography, Peter Taylor defined ideology as 'a world-view about how societies both do and should work. It is often used as a means of obscuring reality' (Taylor 1993: 331). This deceptively straightforward definition itself obscures the battles over the identification, meaning and use of ideology, within political geography and more generally.

Evolution and Debate: Is Anything Outside of Ideology?

The understanding of ideology in political geography cannot be separated from broader debates over the concept within the discipline of geography as a whole. Within geography, David Harvey took issue with the separation of science and ideology, and the resulting conceptualization of science as in some way neutral and separated from ideology. He argued that 'the use of a particular scientific method is *of necessity* founded in ideology, and that any claim to be ideology free is *of necessity* an ideological claim' (Harvey 2001: 39; emphasis in original). The discussion of ideology was further developed by Derek Gregory (1978). Gregory argued that much geographic research described as science would be better described as ideology, since it was 'unexamined discourse'. He posited a more critical approach to geography; one that problematized discourse (Gregory 1978). These approaches to the definition of ideology in many ways signalled the beginning of a shift in geographical thought, as geographers started to focus on discourse, on the contested nature of knowledge and on the possibility (or indeed impossibility) of objective knowledge.

One of the key ways in which political geographers engaged with ideology was to examine it as a source of conflict and contestation. The study of the Cold War is an obvious example (see Figure 11.1). John Agnew suggests that the Cold War represented the replacement of territorial rivalry between imperial powers with an ideological dispute based on political-economic differences (Agnew 2002: 94). Those differences – between the United States with its commitment to capitalism, and the Soviet Union with its commitment to communism – were based on fundamentally opposed ideologies, but had material consequences.

Figure 11.1 Remains of the Berlin Wall, which divided East and West Germany

Initially, the site of contestation was Europe, but by the 1950s and 1960s, conflicts between the US and the USSR were being played out in Asia, Africa and the Caribbean (see Figure 11.2). During this period, John Agnew argues, political geography – in common with many other social sciences – offered little in the way of critique of how ideology influenced Cold War political decisions and actions by both the US and the USSR (Agnew 2002: 94). Instead, political geography generally worked to serve the interests of the major powers and the state (Taylor 2003: 47). As a consequence, initial studies of the Cold War were limited in their scope and intent. However, the revitalization of political geography during the 1960s and 1970s led to a renewed interest in the Cold War and resultant conflicts, both within Europe and in the wider world. In addition, political geographers extended their studies of conflicts over ideologies, focusing on a variety of competing ideologies within states and between states, as well as on the local, international and global scales.

More fundamentally, however, this revitalization has directed explicit attention to the ideologies of political geography itself. This has

Figure 11.2 The Vietnam War Memorial in Washington DC. Vietnam was the site of one such conflict

happened in a variety of ways. One involves an examination of the extent to which the history and development of political geography have been enabled and constrained by its direct association with state power through its provision, in Peter Taylor's words, of 'spatial recipes for the powerful' (Taylor 2003: 47). This has led to a critical reappraisal of the history of political geography: its association with colonialism and imperialism; its links with theories of expansionism in Nazi Germany; and the extent of its involvement in the development and spread of the American Empire in the late twentieth century. (For more information on geography's role in these epochs see Chapters 7 and 9 on geopolitics and colonialism/imperialism respectively.) Another has involved an examination of the objects of political geographic research, particularly the ongoing emphasis on the state as the appropriate scale of analysis. This has included recognition of the state itself as an ideological construction, as well as an admission that political geography extends beyond the state, to individuals, cities, regions, institutions and networks. Yet another relates to ideology in the form of epistemology, especially the ways in which some forms of knowledge have been privileged over others in the construction of political geography (see, for example, Ó Tuathail 1996b; Staeheli et al. 2004). As a consequence, the understanding of appropriate sources for political geography has been considerably expanded.

A variety of recent survey texts suggest that the attention to ideology within political geography as a discipline was facilitated by a number of factors (see, for example, Agnew 2002; Agnew et al. 2003b). The first was the gradual erosion of the Cold War mentality from the period of the Vietnam War onwards. The second was the opening up of political geography, and indeed geography more generally, to so-called 'small p' concerns. Indeed, people from a significantly wider range of social and spatial contexts began to enter the discipline of geography in the 1980s and many were drawn to study the social sciences as a result of social upheaval and protests. The third was the influence of a broad range of new theoretical perspectives, including postmodernism, poststructuralism and postcolonialism (e.g. Gregory and Pred 2006; Ó Tuathail 1996c). Agnew et al. suggest that these factors 'have produced a contemporary political geography that is dynamic and diverse ... distinguished by the critical nature of the questions it asks and the themes it pursues' (Agnew et al. 2003b: 5). The question of ideology in political geography is well addressed in a special issue of *Political Geography* (Cox 2003). This issue was framed around two key questions: theory in political geography; and the meaning of the political in political geography. From the contributions, it appeared that many believed that the political in political geography had been too narrowly defined in the past, focusing on the state rather than more broadly on how power works on different scales. Contributors argued for a political geography that focused on the interconnections of these various scales, for example between the household and the state (England 2003). As Mamadouh suggested in a summary of the debate, 'Political geography needs not closure, in terms of determining which topics and themes should be addressed and in which places, but clarification about specific tools, both concepts and methodologies' (Mamadouh 2003: 672).

However, just as the dominant ideology of political geography has come into focus, challenged by those who are concerned with developing a different kind of normative political geography (Agnew 2002: 164–178), broader events challenge this move towards critical engagement, heterogeneity and inclusivity. In particular, the 9/11, 2001 attacks on the USA led to a renewed engagement by political geographers with matters of concern to the state. This state-centred approach to politics is evident in publications and research projects that direct the focus back to state borders, state boundaries, and the protection of states (see, for example, Cutter et al. 2003). At the same time, other geographers continue to ask critical questions about the role of geography and the role of the state in the ongoing 'War on Terror' (Bialasiewicz et al. 2007; Ettlinger and Bosco 2004; Gregory 2004; Gregory and Pred 2006). In short, debates and contests over the ideologies of political geography continue, and are unlikely to be resolved.

Case Study: Veiling in Turkey

The question of ideology as a site of contestation is obvious in a variety of recent articles on the political geography and geopolitics of religion. This has taken on increased urgency in the advent of the War on Terror, often constructed as a clash between western (read Christian) and eastern (read Muslim) civilizations. A recent issue of *Geopolitics* dealt explicitly with this topic. John Agnew observed, in the introduction to this issue, that 'religion is the emerging political language of the time' (2006: 183). This is as evident at the local level, in the everyday religious practices of people around the world, as it is at the global level in the replacement of a Cold War ideology with a religious ideology as a call to arms.

An example of conflict over ideology is provided by Anna Secor in her study of veiling and urban space in Istanbul (Secor 2002). Secor shows how veiling – the covering of a woman's hair, and possibly face and body – has been an ongoing site of struggle in Turkey. When the Turkish state was formed in 1923, its leaders attempted to steer the country away from an Ottoman-Islamic past, towards a more western and secular future. Women were encouraged to remove the veil, adopt western dress and work alongside men to build the new nation (Secor 2002: 9). Secor points out that this did not lead to a wholesale change: many women continued the practice of veiling, though others, particularly in urban areas, uncovered their heads. While there was no official law against veiling, the practice was prohibited in particular spaces, such as university campuses, courts and Parliament (Secor 2002: 9–10).

Though the Turkish state is secular, the majority of its population is Muslim. Since the 1990s, an Islamist political party has been growing in popularity. Islamist politicians claim that restrictions on veiling inhibit women's freedom, while Islamist women claim that bans on

Figure 11.3 Istanbul University

veiling in places of education discriminate against devout women who wish to participate in higher education (Secor 2002). A re-veiling movement has emerged, with women covering their heads on university campuses as an act of protest (see Figure 11.3). Similar 'new veiling' movements are emerging in other Muslim communities (Secor 2002: 9), for example in Egypt and France.

Secor's article adds to a growing body of literature on Turkey as a site of contestation over ideology: both within Turkey, between secular and religious ideologies (Jefferson West 2006; Secor 2001) and in broader debates over Turkey's possible membership of the European Union. It also adds to a growing body of literature on the geographies of Islam (Aitchison et al. 2007; Falah and Nagel 2005).

The concept of ideology is thus important in political geography in two ways: in the study of ideologies as a source of conflict, contestation and, ultimately, place-making; and in the study of the ideologies of political geography itself. Both continue to be influenced by broader geographical and political contexts and events.

KEY POINTS

- The meaning of ideology has been contested and altered since it first appeared in the English language in the eighteenth century. Within political geography, debates over the meaning of ideology mirror broader debates within the discipline of geography.
- Political geographers have paid particular attention to ideology as a source of conflict, for example by studying the Cold War.
- More recently, political geographers have paid attention to the ideologies of political geography, particularly its state-centred approach. These critical geographers have argued for a broader understanding of politics and epistemology within political geography.

FURTHER READING

Agnew, J. (2002) *Making Political Geography*. London: Arnold.

Cox, K. (ed.) (2003) Forum: 'Political geography in question', *Political Geography*, 22(6): 599–675.

Eagleton, T. (1991) *Ideology: An Introduction*. London and New York: Verso.

Gregory, D. & Pred, A. (eds) (2006) *Violent Geographies: Fear, Terror and Political Violence*. New York: Routledge.

Staeheli, L., Kofman, E. & Peake, L. (eds) (2004) *Mapping Women, Making Politics: Feminist Perspectives on Political Geography*. New York: Routledge.

12 SOCIALISM

Mary Gilmartin

Definition: A Concept with Economic and Social Elements

Socialism, according to Joseph Schumpeter, is a system where 'the control over means of production and over production itself is vested in a central authority' and where, as a matter of principle, 'the economic affairs of society belong to the public and not to the private sphere' (1992: 167). In this way, socialism is defined in opposition to capitalism, where the means of production are privately owned. In his definition, Schumpeter highlights the economic base of socialism: this suggests that all forms of social relations are based on economic relations. Geoff Eley suggests that 'socialism began as the ambition to abolish capitalism, to build an egalitarian democracy from the wealth that capitalism endowed' (2002: 483). Eley argues that this original impetus was tempered by a series of setbacks and defeats, so many socialists focused on more modest aims. These included social citizenship, social justice, rights at work and the insistence on democracy. As a consequence, working definitions of socialism now often highlight social equality and inequality, and a desire for human liberation that extends beyond the economic sphere.

There are many disagreements over the meaning of socialism and many forms of socialist thought and practice. A key distinction is between utopian socialism and democratic socialism. Utopian socialism wanted to move away from the influence of the state and advocated small-scale experimental communities and radical gender politics. Democratic socialism sees a role for the state, as long as it is run democratically. Other forms of socialism include international socialism, revolutionary socialism, state socialism and feminist socialism. International socialism attempted to create a strong international working-class solidarity. It was exemplified by the First and Second

Internationals, international socialist organizations founded in 1864 and 1889 respectively. It continues today with 'new labour internationalism' that attempts to challenge 'multinational capitalism at a transnational scale' (Blunt and Wills 2000: 68). Revolutionary socialism suggests that social change in society – in particular, a move towards public ownership of the means of production – should happen through revolution and the overthrow of capitalism, rather than as a gradual process. Revolutionary socialism is often associated with communism. State socialism advocates that the means of production are owned by the state. Feminist socialism argues that gender as well as class accounts for the subordination of women (Blunt and Wills 2000: 110). Blunt and Wills argue that anarchism also emerged from nineteenth-century socialist thought and that it represents a contemporary alternative to Marxism (2000: 1–39).

It is important to make a distinction between socialism and Marxism, though the two terms are often used interchangeably. Marxism is a body of thought that has emerged from the work of Karl Marx and is concerned with the overthrow of capitalism as a system of societal organization. Marx, together with his collaborator, Friedrich Engels, believed that change occurs because of conflict and opposition between the different elements of a system. Those conflicts occur between different classes of society, as a result of what Marx and Engels saw as the inevitable economic crises of capitalism. Marxian thought identifies the working class – those who do not have control over the means of production – as central to social change. Marx and Engels also identified a new type of social organization to replace capitalism – communism. They argued that communist society would abolish private property (and thus class) and socialize wealth: 'in Communist society, accumulated labour is but a means to widen, to enrich, to promote the existence of the labourer' (Marx and Engels 1848, in Blunt and Wills 2000: 53). A variety of forms of Marxism have been defined. Blunt and Wills, for example, distinguish between classical and western Marxism, where western Marxists were influenced by thinkers other than Marx, and reacted in particular against Stalinism and fascism. The variety in forms of both Marxism and socialism makes it difficult to draw clear and unambiguous boundaries between them. However, at a fundamental level Marxism believes the working class is the key agent for social change, a belief that is not always so strongly held by all forms of socialism.

Evolution and Debate: A Concept on the Margins

The ways in which socialism has been understood and researched within geography more generally, and within political geography specifically, are varied. Broader histories of geographic thought suggest that Marxism was embraced by the discipline of geography from the 1970s onwards, building on the earlier work of geographers interested in socially relevant and radical approaches to the discipline and society (see Peet 1998: 67–111). The journal *Antipode*, established in 1969, was and continues to be a key vehicle for the dissemination of these ideas (see Figure 12.1). In relation to geographers' engagement with Marxism, Richard Peet comments that Anglo-American Marxist geographers 'were preoccupied with deriving analytical concepts and categories from the corpus of Marx's work and applying them to practical issues of immediate concern' (1998: 93). Some of the key works in this period included Neil Smith's *Uneven Development* (1984), as well as *Limits to Capital* and *Social Justice and the City* by David Harvey. In his work Smith argued that uneven development 'is both the product and the geographical premise of capitalist development' (Smith 1984: 155): both a reaction to and a consequence of the movement of capital in search of profit. In *Limits to Capital* David Harvey described the spatial dimension to capital accumulation as the 'spatial fix' – by this, he means the way in which capitalists can deal with falling profits in one part of the world by reinvesting their capital in another part of the world (for example, moving manufacturing industry from a high wage to a lower wage economy). In this way, Harvey argues, space is fundamental to the ways in which capitalism works (Harvey 1982), and he insists on the importance of a historical-geographical materialism that considers both space and time in its analysis of capitalism (2000: 15). This body of work may be broadly described as having a political economy focus. (For more information on political economy see Chapter 10.)

However, despite the wide reach of Marxist geography and its insistence on the relationship between politics and economics, Peter Taylor writes that it initially had little influence in political geography (Taylor 2003). He observes that political geography has traditionally been conservative in orientation, and that it had 'little or no engagement with the new radical geography that emerged in the 1960s' (2003: 48). Taylor suggests that the influence of Marxism on political geography began

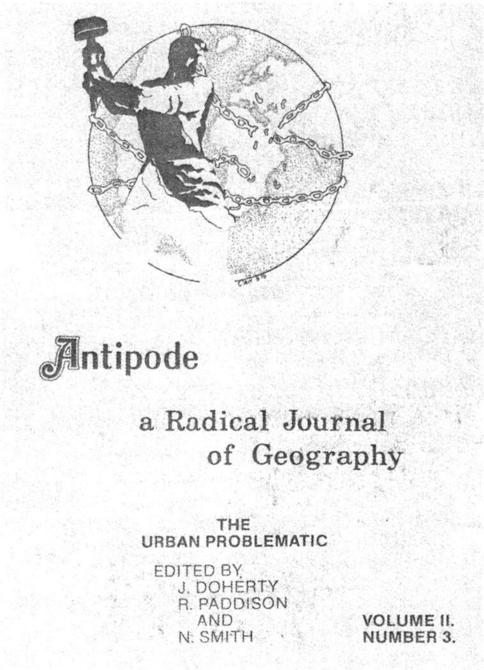

Figure 12.1 Early *Antipode* cover

with a new focus on the state as an integral part of the capitalist system. Some political geographers began to argue that political processes could not be studied separately from economic processes. The advent of world systems political geography was an attempt to use insights from Marxist thought, in particular the work of Immanuel Wallerstein, to study spatial organization and social change. Wallerstein argued that social change should be studied as global rather than within the confines of the state, and from a long-term historical perspective. His ideas were taken up by political geographers such as Peter Taylor, and used to create a radical theoretical framework for political geography, though recent texts (for example, Jones et al. 2004b: 10) suggest that the world systems approach is not widely used by political geographers. Political geographers have also used the *régulation* approach to political economy and the state. This argues that continued capital accumulation requires a range of political, social and cultural supports, for example from the state and other institutions. These supports are

Figure 12.2 This statue of Tito in Valenje, Slovenia, is one of the few remaining statues of Tito in former Yugoslavia. Many monuments to socialism and communism were destroyed, in Yugoslavia and elsewhere. (Photo by and courtesy of Mary Gilmartin)

particularly necessary at times of crisis (see Jones et al. 2004b, Chapter 4, for a good overview).

In recent years, political geographers have devoted their attention to socialism by studying the geographies of post-socialist and post-communist states, regions and cities (see Figure 12.2). This mirrors an earlier engagement by political geographers with the geographies of the Cold War. Recent research has focused on the process of transformation from socialism and communism to a post-socialist or post-communist society. This research is wide-ranging, with an emphasis on topics as varied as urban planning and urban space (Pavlovskaya and Hanson 2001), environment and resources (Pavlínek and Pickles 2000), national identity (Young and Light 2001), and democratization, studied at scales ranging from the local to the international. (For more information on democratization see Chapter 4.) However, much of this research is carried out by geographers who would not necessarily describe themselves as political geographers. In this way, studies of socialism, communism and their aftermath are as likely to focus on cultural or economic transformations as on political transformations. Nonetheless, studies of post-socialist and post-communist societies provide one of the means by which political geographers continue to engage implicitly with the concept of socialism.

Readers interested in finding a definition for socialism in contemporary political geography texts could be forgiven for thinking that socialism no longer existed. Few recent survey texts on political geography make explicit reference to the concept, let alone attempt to define it (see Cox 2002 for an exception). Instead, references to socialism are couched in discussions of capitalism, neoliberalism, the Left, critical geography and, in a few rare instances, Marxist geography. There has been, Blunt and Wills suggest, a decline in interest in Marxism in recent years (2000: 42). Since Marxism is often seen as interchangeable with communism and socialism, this has served to marginalize the position and study of socialism within political geography. Despite its apparent absence from political geography, it is clear that a variety of radical approaches to geography – including socialism – continue to be important and remain the focus of heated debate within the discipline.

One such debate has been occurring in the pages of *Antipode* and *Transactions*, where the meaning of 'Left' as articulated by Amin and Thrift has been taken to task by, among others, Neil Smith, David Harvey and Jane Wills (Amin and Thrift 2005, 2007; Harvey 2006; Smith 2005a; Wills 2006). While the various protagonists agree that inequality, injustice, ethics, rights, power and politics are important issues, they fundamentally disagree on how to address these issues. Smith, Harvey and Wills all insist that capital and class have to be the primary focus of 'Left' geographies, while Amin and Thrift argue for less emphasis on capitalism and a more pluralistic approach. How their debates might be translated into political geography as yet remains unclear.

Political geography has also been relatively silent on the recent move to the left in Latin America, as evidenced by recent victories for left-wing presidential candidates in Venezuela, Chile, Brazil, Argentina, Uruguay, Bolivia and Peru (Cleary 2006). The extent to which this is a regional or international trend, and its implications for political geography more broadly, are questions yet to be addressed in print by political geographers.

Case Studies: Post-socialism in Eastern Europe and Imagined Alternatives

There are two bodies of research that I wish to highlight in order to give a snapshop of research on socialism. The first addresses the issue of socialism through a focus on post-socialist societies. The second outlines a socialist alternative to contemporary capitalism.

In relation to post-socialism, the work of Alison Stenning focuses on the changing everyday geographies of people living in such societies. Stenning's work looks at Nowa Huta, a district of Kraków in Poland (Stenning 2000, 2003, 2005). Nowa Huta was founded in 1949, and centred on a new steelworks plant. For many decades, the plant provided secure employment, as well as a wide range of local social and cultural facilities and services. With the collapse of socialism and consequent economic and social restructuring, these facilities and services were often pared back. For example, an extensive and cheap public transport system existed under socialism, but under post-socialism, public transport has declined in favour of private car ownership. Similarly, under socialism, the steelworks provided subsidized excursions and travel, as well as access to company holiday resorts, for its workers and their families. Now, there are no guarantees of travel, and while some residents of Nowa Huta have the resources to travel widely, many others are 'socially and spatially entrapped' (Stenning 2005: 121). The network of local shops that existed under socialism remains, but the shops have been complemented by supermarkets and hypermarkets, and by the emergence of open air markets offering a wide array of goods at low prices. Stenning's study makes the point that post-socialism is 'marked by a contradictory widening of horizons and the shrinking of lifeworlds' (Stenning 2005, 124): she sees this as related to a range of factors, including global economic restructuring as well as the historic experience of socialism. Her work is supported by a recent special issue of *Geoforum*, which provides insight into the variety of ways in which socialism and communism worked and are reworked (Herrschel 2007). For example, Lindner highlights the micro-geographies of power and accumulation that emerged in rural Russia following the decision to privatize agricultural collectives (Lindner 2007), while Lintz et al. discuss the different problems faced by industrial cities and regions in former socialist and communist states, such as high unemployment and ecological devastation (Lintz et al. 2007). Through these and other case studies (e.g. Smith and Stenning 2006), few of which are carried out by political geographers, our understanding of socialism in practice is expanded and made more nuanced.

Few geographers make explicit reference to socialist alternatives. An exception is David Harvey, whose *Spaces of Hope* (2000) considers utopian socialism, and in the book's concluding appendix he outlines a personal vision of a socialist utopia. Harvey's vision of this new society is organized around small units of habitation, of between 20 and 30 adults with associated children, who make collective decisions about how to live and how to organize. This unit – a hearth – is also part of a variety of other units. Harvey suggests that the largest political unit, made up of between 4,000 and 10,000 hearths, should have no more than three million people. In this utopian society, private property would be abolished, and government is direct and elected rather than the preserve of a bureaucratic elite. This vision of a socialist society of the future has been critiqued or ignored by many within geography, but it is remarkable for its assertion of a socialist-inspired future by one of the key thinkers of contemporary geography.

KEY POINTS

- Socialism is often defined in opposition to capitalism. Under socialism, the means of production are publicly owned, in contrast to capitalism, where they are privately owned.
- There is a variety of forms of socialism, of which Marxism is just one. Marxism has been particularly influential within geography generally, though not within political geography.
- The study of post-socialist societies has been a central focus of recent work within political geography.

FURTHER READING

Blunt, A. and Wills, J. (2000) *Dissident Geographies: An Introduction to Radical Ideas and Practice*. Harlow: Pearson Education.

Cox, K. (2002) *Political Geography: Territory, State and Society*. Oxford: Blackwell.

Eley, G. (2002) *Forging Democracy: The History of the Left in Europe 1850–2000*. New York: Oxford University Press.

Harvey, D. (2000) *Spaces of Hope*. Berkeley: University of California Press.

Stenning, A. (2005) 'Post-socialism and the changing geographies of the everyday in Poland', *Transactions of the Institute of British Geographers*, 30(1): 113–127.

13 NEOLIBERALISM

Carolyn Gallaher

Definition: A Return to Laissez-Faire

The term neoliberalism is often used to describe a set of economic policy initiatives that came to prominence in the early 1980s. These policies represent an updated version of laissez-faire economics, which dominated economic thinking in the nineteenth and early twentieth centuries. Like its predecessor, neoliberalism is based on the assumption that the economy works best when left alone. Tariffs, quotas, subsidies and other forms of government intervention in the economy are viewed as hindrances to economic growth. Likewise, state ownership of the means of production is discouraged as an inefficient mechanism for ensuring a wide access to resources.

Neoliberalism is known by a variety of terms. In the United States it is frequently referred to as the Washington Consensus (Naím 2000). In Great Britain commentators often call it Thatcherism, after the Prime Minister, Margaret Thatcher, who instituted the country's first neoliberal policies (Harvey 2000). In other parts of Europe it is labelled liberal-productivism or monetarism (Lipietz 1992). In 1980 only a few states had adopted neoliberal policies. Today most states in the world employ a neoliberal approach. Institutions like the International Monetary Fund, the G8 and the World Bank were vital in spreading neoliberal policies to developing countries and post-communist states in the eastern bloc (Ould-Mey 1996; Stiglitz 2003). These organizations required developing countries to adjust their economies along neoliberal lines in order to receive development loans. The global reach of neoliberalism has led some scholars (Agnew and Corbridge 1995) to define it as a hegemonic system. (For more information on hegemony see Chapter 5.)

While neoliberalism is generally defined as a set of economic policies, geographers argue that it is better understood as an ideology. (See Chapter 11 for more information on ideology.) It is a world view about how wider society *should* be organized (Smith 2005b). In particular, neoliberalism is a repudiation of the Keynesian state and the economic and social policies that came with it (Smith 2005b; Stiglitz 2003). The French economist Alain Lipietz characterizes Keynesianism as 'a grand compromise' between the Left and Right, labour and capital. Capitalists could mass-produce their goods and use Taylorist methods to increase productivity,[1] while workers were assured that productivity gains would be matched by higher wages. The state's job was to keep the overall system stable. The state could, for example, raise interest rates to prevent inflation. The state was also expected to intervene if the compromise threatened to fall apart. In the United States the National Labor Relations Board was established to mediate between business owners and workers threatening to strike. The group's stated mission was to avoid work stoppages by negotiating compromises between labour and capital.

Neoliberal ideology represents a critique of the economic *and* social bases of the Keynesian system. In matters of economic policy, neoliberals argue that Keynesian policies stymied economic growth. They suggest, for example, that high tax rates discourage capitalists from reinvesting profits. Likewise, subsidies hinder innovation outside subsidized sectors of the economy (Naím 2000). In the social realm, neoliberals depicted Keynesian welfare policies as well-meaning, but misguided (Friedman 1962). They argued, for example, that welfare payments rewarded those who chose to drop out of the job market (Murray 1994). Similarly, subsidies for peasant agriculture in the developing world allowed farmers to avoid becoming 'productive' members of society.

While neoliberalism represented a critique of the Keynesian state, and by extension the left-of-centre parties that tended to support its principles, neoliberal ideology is not the sole province of right-of-centre political parties or coalitions. Indeed, while the conservative leaders Ronald Reagan and Margaret Thatcher are often seen as the architects of neoliberalism (see Figure 13.1), many of the ideology's key tenets were carried to fruition by left-of-centre leaders such as Bill Clinton in the US and Tony Blair in Great Britain (Smith 2005a). In his 1996 State of the Union address, President Clinton famously declared that 'the era of big government [was] over', and he made good on the promise by changing the US welfare system to limit the duration of time that recipients could receive welfare.

Figure 13.1 Neoliberal Architects: Ronald Reagan and Margaret Thatcher 1981. (Photo courtesy of the Ronald Reagan Library)

Evolution and Debate: From Margin to Centre

Intellectual Foundations

There is no one intellectual 'father' or 'mother' of neoliberal thought. Several thinkers, in Europe and the United States, developed ideas that would eventually gel into a cohesive ideology. In Europe, Friedrich von Hayek led the charge when he formed the Mont Pelerin Society in 1947 to advocate free market ideals. The society's opening salvo declared that western civilization was in danger from policies that undermined individual freedoms. In the United States Walter Rostow and Milton Friedman were important advocates, as was the Department of Economics at the University of Chicago (Smith 2005b). Until the late 1970s, however, free market ideals remained the province of intellectuals, gaining little traction in policy circles.

Material Foundations?

Accounts of the rise of neoliberalism vary, depending in part on theoretical orientation. Standard or mainstream views depict neoliberalism as an innovation of capitalism. Halal (1986), for example, depicts Keynesianism as an aberration in capitalist development, and neoliberalism as a mid-course correction. Political economists (Harvey 1990, 2007; Smith 2005b) tend to view the rise of neoliberalism as a response to a capitalist crisis of accumulation. (For more information on political economy see Chapter 10.) David Harvey, for example, connects the rise of neoliberalism to a crisis of overaccumulation in the 1970s. In simple terms overaccumulation is 'a condition in which idle capital and idle labour ... exist side by side with no apparent way to bring [them] together' (Harvey 1990: 180). Societies experiencing a crisis of overaccumulation are marked by excess commodities, high unemployment and surplus budgets. For Marxists, like Harvey, overaccumulation is a recurring crisis within any capitalist mode of production and cannot be eradicated, only managed.

American and Western European economies, which were based on a Fordist mode of production, were able to avoid the problems of overaccumulation for almost a quarter of a century. Indeed, the post-World War II boom is often regarded as the longest in modern capitalist history. Harvey argues that Fordist economies were able to avoid crisis through 'spatial and temporal displacement' (Harvey 1990: 185). Spatially, capitalists prevented diminishing rates of return (on profits) by creating 'new geographical centres of accumulation'. These new centres included domestic peripheries, such as the southern states in the US, and international ones, such as newly industrializing countries (NICs) in Asia. Temporally, western economies used debt to create new financial products, thereby absorbing excess capital.

In time, however, these strategies began to sink under their own weight. New centres of accumulation became competitive and forced older, more established centres of industry to contract. Increasing debt loads led governments to print more money, contributing to inflationary pressure. Likewise, the 1973 OPEC oil embargo created an energy crisis when the price of a barrel of oil increased 470 per cent in one year (Silke 1973). To placate oil-dependent governments the largest OPEC producers agreed to invest their newfound wealth in western banks

(Harvey 2007). While banks were happy with the sizeable deposits, they also needed borrowers big enough to absorb their newfound capital. Developing countries, many of them newly independent and short on capital, were obvious candidates for absorbing the petrodollars pouring in from the Middle East, and new loan products were quickly developed. However, most of these loans were in dollars with adjustable interest rates, making developing countries vulnerable to changes in US interest rates (Harvey 2007).

That vulnerability was fully exposed in 1979 when Paul Volcker, then chairman of the US Federal Reserve Board, instituted the first in a series of sharp interest rate hikes to rein in domestic inflation. As David Harvey (2007) notes, Volcker's interest rate hikes represented an opportunity to put neoliberal ideas into practice. The Keynesian emphasis on full employment was abandoned in favour of policies designed to inhibit inflation at all costs, even if doing so triggered a recession. In western economies the sharp interest rate hikes hastened deindustrialization as the recession drove companies to shut down factories and force concessions from the unions in others. In the US it also ushered in the farm crisis as debt-laden farmers (who were domestic recipients of banks' newly acquired petrodollars) succumbed to foreclosure (Dyer 1998; Staten 1987). In the developing world debt payments skyrockected and countries like Mexico and Brazil came close to defaulting. Their bail-outs, constructed by Reagan and Thatcher appointees at the IMF, required neoliberal adjustments in return.

Geographic Origins

While neoliberalism is usually regarded as an American and European export to the developing world, it bears mentioning that neoliberal policies were first instituted not in the US or Britain, but in Chile and Argentina *before* the debt crisis began (Harvey 2007). The events that marked their application were bloody. In Chile General Augusto Pinochet overthrew Chile's democratically elected President, Salvador Allende, in 1973. Allende, who had run on a socialist platform, was depicted as a communist menace. Pinochet's military junta promptly implemented a neoliberal economic plan for the country, eliminating protectionist policies and dismantling many of the state's social service programmes. It also violently repressed domestic opposition to the changes. A similar chain of events occurred in Argentina three years later (1976) when Isabel Peron was removed from office in a military coup. The military junta then implemented a number of market reforms to open up the Argentine economy. And, as in Chile, domestic opponents were ruthlessly suppressed.

While neoliberalism began outside the US, rather than in it, it is unlikely it would have spread globally from Chile and Argentina without first being adopted by the US and other G8 members (Harvey 2007). Western leaders controlled the purse strings of international lenders like the World Bank and International Monetary fund and were able to use these organizations to compel cash-hungry governments in the developing world to adopt otherwise domestically unpopular policies in exchange for foreign capital.

Although dictators and military juntas were able to implement neoliberal policies by force, democratic nations had to manufacture sufficient consent among the voting populace to enact them. In developed nations, the stagflation of the 1970s and the recession that followed allowed leaders like Ronald Reagan and Margaret Thatcher to depict neoliberal policies as unavoidable (Harvey 2007). Margaret Thatcher succinctly captured this view when she scolded her domestic opponents with the mantra 'there is no alternative' (Harvey 2000). This is not to suggest, of course, that neoliberal reforms inspired little opposition; domestic resistance was initially fierce on both sides of the Atlantic. However, free market advocates were able to maintain the momentum established by Reagan and Thatcher.

Discursively, neoliberal boosters tied free market ideals to notions of personal liberty and freedom (Harvey 2007). They also created an institutional support base for neoliberal ideology, building conservative think tanks, funding free market research by university professors and later supporting political candidates who espoused their ideas. In the 1980s in the US, for example, conservative think tanks like the Heritage Foundation and the American Enterprise Institute were established to pursue pro-market research (Harvey 2007). Corporate seed money allowed these groups to develop, publish and disseminate free market policy prescriptions to members of Congress. Political parties were also co-opted as free market interest groups threatened to withhold donations to political parties that opposed the 'reforms'. The growing influence of big money in elections was especially difficult for the Democratic Party, which was left with the unenviable choice of trying to win pro-labour campaigns against better-funded opponents, or adopting neoliberal positions to obtain the cash necessary to remain competitive (Harvey 2007). The rise of the 'pro-market' Democratic Leadership

Conference in the US Democratic Party was a direct result of these influences. So, too, was the rise of so called 'New Labour' in Britain.

Neoliberal leaders also used union-busting to thwart opposition from workers. In 1981 Ronald Reagan signalled a get tough approach to labour opposition when he fired members of the Professional Air Traffic Controllers Association. Labour would also suffer as its share of the workforce employed in the manufacturing sector, the most unionized sector of the economy, declined with deindustrialization.

Case Study: Bolivia's Water War

Its critics argue that neoliberalism represents a transfer of wealth from the many to the few. This process is often referred to as accumulation by dispossession or primitive accumulation because it relies on a transfer of assets rather than the creation of new capital (Harvey 2007). Accumulation by dispossession can occur through a variety of mechanisms, including the restriction of credit, harsh bankruptcy laws and coercion. The privatization of state-owned enterprises is another common mechanism. In the developing world, for example, the IMF and the World Bank have required governments to privatize state-owned industries as a condition for receiving loans. World Bank and IMF policies in Bolivia provide a case in point (Assies 2003; Perreault 2006; Spronk and Webber 2007).

Water Privatization in Bolivia

During the 1960s and 1970s the World Bank supported the public ownership of water. At the time, most development economists believed affordable (if not free) access to potable water was a necessary precondition for development (Spronk and Webber 2007). Industrial plants need a modern water infrastructure and workers are less likely to miss work as a result of waterborne illnesses common to places without modern water systems. By the 1990s, the World Bank had shifted its position. The bank argued that water was becoming a scarce resource and that consumers would only curtail consumption if they knew how much water usage actually cost.

In the mid 1990s the World Bank offered Bolivia a development loan to modernize the water systems in the country's biggest cities, La Paz-El Alto[2] and Cochabamba (Figure 13.2). The loans were designed to

Figure 13.2 Bolivia's Major Cities (Map courtesy of the University of Texas Perry-Castañeda Library map collection)

make the cities' water systems attractive enough for privatization. In 1998 the IMF made water privatization a reality when it stipulated that Bolivia must privatize its urban water systems in order to receive an anti-inflation package of $138 million (Frontline 2002). The Bolivian government accepted the loan and put the public utilities that ran the country's water systems up for auction.

While the government's water privatization plan was controversial across the country's urban centres, this was particularly so in Cochabamba where the infrastructure was in poor shape, access to water averaged only four hours a day, and over 40 per cent of residents were not connected to the system (Spronk and Webber 2007). When the city system was put up for auction in 1999 only one entity placed a bid, for $2.5 billion. The bidder, Aguas del Tunari, was a consortium of companies headed by Bechtel, a multinational corporation based in San Francisco, California.

A sticking point for the residents of Cochabamba was the contract between the Bolivian government and Aguas del Tunari. Although the contract stipulated that the consortium would extend water and sewerage services to the entire city of Cochabamba, it also guaranteed the consortium the right to a 13 per cent minimum annual return on its investment. Residents feared the consortium would have to raise water rates sharply to receive the minimum return. Their fears were justified; within the first year alone, the consortium raised water rates by close to 200 per cent for some customers (Perreault 2006).

The passage of Law 2029 by the Bolivian government the following year also angered residents of Cochabamba. The law extended control of the city's water system to include informal water systems that were not directly connected to the city's water delivery infrastructure. In essence, the law stipulated that only private companies could distribute water and that any water being distributed by other means would fall under the control of the private companies in question. This was a particular concern in Cochabamba where upwards of 40 per cent of the population was not connected to the main water system. In areas where residents were not hooked into the water system, a series of customary regulations (known locally as *uso y costumbres*) were in place to govern the right to water. Opponents noted that Law 2029 was written so broadly as to permit the consortium to seize control of individual cisterns if it so wanted (Olivera 2004).

Water privatization in Cochabamba sparked a violent 'water war', with numerous protests and street clashes in the city and in the capital (see Figure 13.3). In April 2000, the government of Bolivia rescinded its contract with Augas del Tunari (Perreault 2006). The company responded by filing a $25 million claim against Bolivia with the International Centre for Settlement of Investment Disputes (ICSID), an arbitration board established by the World Bank (Frontline 2002). In 2006 Aguas del Tunari and the Bolivian government reached a settlement through ICSID. Aguas del Tunari dropped its claim for $25 million and the Bolivian government announced that the contract was revoked 'only because of the civil unrest and the state of emergency in Cochabamba and not because of any act done or not done by the international shareholders of Aguas del Tunari' (Bechtel Corporation 2006).

Figure 13.3 Riot Police in the streets of La Paz, Bolivia

Aguas del Tunari's decision to drop its claim was due in large part to the inordinate share of bad press that Bechtel, the consortium's lead investor, had received in the western press. One story, in 2002 in the *San Francisco Chronicle*, described the conflict as 'David fighting Goliath', noting that Bechtel's revenues for 2002 were more than six times greater than the entire Bolivian budget for the same year (Frontline 2002). American opponents of the ICSID claim also organized a letter and email campaign against Bechtel, demanding that the company drop its claim (Frontline 2002).

Had water privatization in Bolivia worked, residents of Cochabamba would have witnessed a severe form of accumulation by dispossession. Access to a resource every citizen needs would have been transferred from a collective regulatory scheme to a private one, with the restrictions to access this entailed. Potential limits to water struck a visceral chord for many. Water was, in the words of privatization opponents, life (Perreault 2006).

While many commentators rightly drew wider lessons from the events in Cochabamba, geographers took pains to note that the events there also demonstrated the particular, rather than unitary aspects of neoliberalism. At a macro level, Cochabamba illustrates that neoliberalism operates differently across the development divide. As Perreault (2006) argues, water privatization in Bolivia did not follow the standard 'Western' script on neoliberalism in which a state-owned public utility is 'hollowed out'. In Bolivia, water provision was never fully centralized, so privatization did not raise questions about what services the state should provide so much as how resources and access to them should be governed, regardless of the entity in charge of them. And these questions necessarily brought to the fore questions about which values should govern systems of access. In particular, this opposition to water privatization turned on notions of social justice – both procedural and distributive (Perreault 2006). Procedurally, opponents argued that the sale of Cochabamba's gas should have been a transparent and consultative process. In the matter of distribution, they argued that the distribution of water should be equal.

Geographers also note that neoliberalism, and the responses to it, varied within the country. The questions that governed the debates about water privatization were, for example, far different to those that framed the country's attempt to privatize its natural gas and oil deposits (Perreault 2006). Whereas privatization of water turned on questions about the how water should be accessed domestically, questions about the privatization of Bolivia's gas deposits centred on who should consume Bolivia's resources – its domestic population or a foreign one.

KEY POINTS

- Neoliberalism emerged in the 1980s as an updated version of laissez-faire economics.
- Mainstream scholars depict neoliberalism as an innovation of capitalism. Political economists view the rise of neoliberalism as a response to a capitalist crisis of accumulation.
- Augusto Pinochet, Ronald Reagan and Margaret Thatcher are the three world leaders most often associated with neoliberalism.
- Critics argue that neoliberalism represents a transfer of wealth from the many to the few. This process is often referred to as accumulation by dispossession because it relies on a transfer of assets rather than the creation of new capital.

NOTE

1 In *The Principles of Scientific Management* Frederick Taylor (1911) laid out a system for increasing industrial efficiency and output. Rather than one worker doing the several tasks necessary to make something, Taylor argued that production would be faster if it was broken into discrete tasks with a different worker responsible for each task. Taylor's ideas formed the basis of the assembly line, where each worker added a piece to the good being produced.

2 El Alto is a suburb of La Paz.

FURTHER READING

Friedman, M. (1962) *Capitalism and Freedom*. Chicago: University of Chicago Press.

Harvey, D. (2007) *A Brief History of Neoliberalism*. Oxford: Oxford University Press.

Naím, M. (2000) 'Washington consensus or Washington confusion?' *Foreign Policy*, 118: 87–103.

14 GLOBALIZATION

Alison Mountz

Definition: Connecting Many Worlds

Globalization involves the intensification of flows across the globe. These flows include people, goods, ideas, trends, services and money across the boundaries of localities, nation-states and regions. The result is more intense and deeper connections between places. The immediacy of these linkages is associated with advances in technology: the telegraph, the telephone, and eventually the facsimile machine, Internet and aeroplane. In particular, life since the inception of the Internet in the 1990s has proved dramatically different from life before it. Whereas people once wrote letters, messages recorded by hand that would be delivered at whatever pace possible (by horse, carriage, and eventually automobile and plane), today we can communicate via 'instant message', talk in a 'live' chatroom or watch a 'webcam' of another place, knowing instantaneously what is happening in multiple sites, perhaps even among people in places situated on opposite sides of the earth. These connections transcend time and space, connecting people, ideas, information and resources across vast geographical distances.

In short, globalization is the functional integration of economic, political, social and cultural processes across space, and most notably, international boundaries (Dicken 2003: 12). Whereas internationalization is said to be at least 500 years old and to include the spread of industrialization, globalization involving the production of goods and services across borders began only in the second half of the twentieth century, specifically in the 1970s when the profitability of producing goods in one country – such as the United States – fell (Dicken 2003: 9–10). Companies looked for cheaper places to produce goods and services, especially labour intensive ones, where they would pay workers lower wages.

Globalization involves what geographer David Harvey (1990) calls 'time-space compression'. Time-space compression is the idea that connections around the world enabled by modern technologies have had the effect of compressing space. In other words, it takes less time to traverse space. In the 1850s, Karl Marx called this the 'annihilation of space by time'. Advances in technologies of communication and transportation have since accelerated time-space compression such that people around the world can be in communication with one another simultaneously in 'real time' and can travel or exchange goods very quickly, overnight.

These changes have also led to the 'new international division of labor' (Fröebel et al. 1980). Multinational or transnational corporations functioned as the principal engines of globalization and sought out 'greenfield sites', or new production sites where labour was abundant, obedient, young, and untainted by histories of unionization and labour protest. Most often, they found this labour in the countries of the global south, with much of the workforce constituted by women of colour.

As Peter Dicken (2003, 2007) notes, 'Every age has its "buzzword" – a word that captures the popular imagination, and becomes so widely used, that its meaning becomes confused.' In his oft-cited text, *Global Shift,* Dicken (2003: 7) traces the use of the term 'globalization', finding that between 1980 and 1984, only 13 publications used the word in titles; 78 did so between 1985 and 1989; but 600 did so between 1992 and 1996. While many authors wrote about globalization in the late 1980s, few agreed on a common definition.

Evolution and Debate: Much to Argue About

Despite globalization having captured the popular and scholarly imagination in the 1980s and 1990s, authors are still debating the characteristics and effects of globalization and the degree to which changes in world cultures, economies and landscapes can be attributed to the empirical phenomenon we call globalization. It follows that, without agreement on its scope, scholars have also disagreed on appropriate theories of globalization. In this section three debates about globalization are discussed.

Globalization literature really began on an economic terrain with an interest in the speed at which capital flowed from one site to another. Economic globalization took off in the late 1970s and early 1980s when production became too expensive in western industrial centres. As Harvey (1990) argues, western economies faced a classic crisis of overaccumulation and needed a 'spatial fix' to shore up their lagging profit margins. (For more information on capitalist crises of overaccumulation see Chapter 13 on neoliberalism.) The new 'international division of labor' (Fröebel et al. 1980) afforded and exploited new opportunities for profit abroad. The shift to production abroad was accompanied by another shift: from the mass production and expensive inventories of Fordism to the cheaper and 'just-in-time' processes of flexible production. The latter relied on cross-regional or even global production chains wherein pieces of large products might be assembled in different places. Today, for example, a car can be produced with steel from China and spark plugs from Singapore, and then assembled in Toluca, Mexico. Computers, too, often contain parts manufactured in multiple countries.

Corporations grew larger and more centralized and, with the advances in technology mentioned above, they were able to globalize production during the latter half of the twentieth century. They would establish headquarters in large world cities and locate manufacturing and even 'back office' functions in other smaller, often more affordable places to do business and move goods (Figure 14.1). Banks became international and consolidated into fewer and fewer large, centralized corporations. Media conglomerates similarly consolidated. Even the corporations that build and manage prisons consolidated into a small number of key players.

The first major debate about globalization stems from the economic view of the topic and involves the role of nation-states in the global economy (Dicken 2003). The empirical shifts noted above prompted a group of authors that Peter Dicken labels 'hyperglobalists' (e.g. Ohmae 1995) to speculate that the nation-state is dead, that it has lost its power and its role in this new hyperglobal environment where business transcends national boundaries. The scholars Dicken labels the 'skeptics' (see Held et al. 1999), on the other hand, were not so fast to declare the death of the nation-state. Dicken himself, for example, says that the state remains an important actor within the global economy, performing the role of economic and political regulator.

While some debated the role of the nation-state, a second group debated more cultural issues, and specifically the threat of homogenization. If McDonald's and Starbucks could operate in nearly the same fashion everywhere, would a 'global culture' based largely on American tastes and norms take hold, penetrate and eradicate the very particulars of local cultures and languages (see Figure 14.2). This second debate about globalization, staged in the 1990s, asked whether globalization would result in 'McDonaldization' – as in the homogenization of

Figure 14.1 Georgian President Mikheil Saakashvili at the opening ceremony for a Pepsi bottling plant. In developing economies, the opening of a multinational corporation subsidiary is an important event.

cultures around the globe – or 'hybridization', meaning a blending of local cultures and languages in hybrid cultural practices and spaces. Benjamin Barber (1996) argues, for example, that globalization and its attendant cultural homogenization, which he labels McWorld, is fostering a violent backlash of tribalisms. Jihadist movements, which call for traditional social and economic norms, are the most pressing example. In contrast, scholars like Featherstone (1990) argue that local cultures are robust enough to incorporate new elements without disappearing altogether. Indeed, the world will more likely see the diffusion of hybrid cultures.

Meanwhile, the provocative language of penetration and eradication so ubiquitous in discourse about globalization prompted a third debate, this one led by feminist scholars (e.g. Kofman 1996; Marchand and

Figure 14.2 McDonald's restaurant in Seoul, Korea

Runyan 2000; Nagar et al. 2002). Feminists argue that globalization scholars often write the *global* and *local* into hierarchical frames (Freeman, 2001), with phenomena categorized as macro-level economic processes weighing more heavily than those classified as micro-level or cultural. That is, globalization becomes the formative backdrop to life's daily minutiae. Carla Freeman (2001: 1008) argues that 'not only has globalization *theory* been gendered masculine but the very processes defining globalization itself – the spatial reorganization of production across national borders and a vast acceleration in the global circulation of capital, goods, labour, and ideas ... – are implicitly ascribed a masculine gender". Sue Roberts (2004) asks, 'Why does the global seem to preclude gender?" She observes the repeated coding of the local as feminine and the global as masculine (see also Freeman 2001).

Finding much of the discussion about globalization to be disembodied, feminists seek to reclaim alternative and embodied sites, voices and ways of knowing the world. Women workers, for example, often experience globalization distinctly from men. Workers at the bottom of the international division of labour have a different perspective on the global economy than those at the top. Just as much 'contemporary political geography describes a "world without people" or at least a world of abstract, disembodied political subjects' (Staeheli and Kofman 2004: 5), so too is globalization discourse depopulated in most renderings. But if technology is a social process, as Dicken argues, so too is globalization. Unfortunately, people appear belatedly in analyses as messy bodies that spoil the smooth surfaces of roving global capital.

Feminist scholars argued that globalization was simultaneously an intimate series of embodied social relations that include mobility, emotion, materiality, belonging and alienation. They asked whether knowledge production of a different kind was required to re-form dominant discourses of globalization: a politics of living and knowing the global. As Mountz and Hyndman (2006: 447), suggest 'The intimate encompasses not only those entanglements rooted in the everyday, but the subtlety of their interconnectedness to everyday intimacies in other places and times: the rough hands of the woman who labors, the shortness of breath of the child without medication, the softness of the bed on which one sleeps".

Globalization discourses perpetuate the myths that the global and local are somehow separate phenomena and that the global somehow prevails over, constitutes, penetrates the local. Yet the local is inextricable from the global (Mountz and Hyndman 2006). The local and global are neither separate spheres nor bounded subjects. Doreen Massey (1993) long ago debunked this idea, showing the local to be constituted by processes, politics and people that exceed its boundaries. Rather, phenomena occurring at local and global scales co-constitute places like the border, the home and the body that require a relational understanding, 'particularly significant in a transnational age typified by the global traveling of images, sounds, goods, and populations' (Shohat 2001: 1269). As Carla Freeman (2001: 1008–1009) 'demonstrates quite literally ... not only do global processes enact themselves on local ground but local processes and small scale actors might be seen as *the very fabric of globalization*' (emphasis added).

Most feminist analyses of globalization would assume the everyday engagements of women and men, including the ways in which relations of work and play, production and consumption defy any fixed or given scale: they are at once connected to global and local processes, politics and people. 'A gendered analysis of globalization would [also] reveal how inequality is actively produced in the relations between global restructuring and culturally specific productions of gender difference' (Nagar et al.,

2002: 261). A feminist analysis would travel farther and develop 'a broader critique of the social production of difference and the multiple exclusions enacted by dominant groups and institutions' (Pratt 2004: 84).

Globalization has bodily expressions (Mountz and Hyndman 2006). Because so much discussion of globalization dwelt in the realms of economic flows and cross-border business, however, the dramatic realm of human migration has often been overlooked (Smith and Guarnizo 1998). Transnational migration is one expression of globalization that has economic, social, cultural, and political dimensions (Basch et al. 1994; Glick-Schiller et al. 1992). Feminists have extensively researched global processes, including the gendered divisions of labour and identities produced by international capital to serve its interests (Marchand and Runyan 2000), as well as the gendered effects of structural adjustment programmes (Lawson 1999).

Case Study: Luisa's Story

Luisa lives outside the city of Oaxaca in southern Mexico where so many men had emigrated to the United States that by the 1990s her village came to be known locally as a village of women (Figure 14.3). Like most households in the town, Luisa's is characterized by absence. Her husband left to work in the service industry in Poughkeepsie, a small city in upstate New York. Unlike other families, however, his has not received routine remittances. Over the years, Luisa has watched her neighbours flourish economically with the support of US remittances. They have enhanced their homes with brick walls, concrete floors and even second-floor additions. They are able to send those children who remain in the village to school and to buy them new clothing.

Luisa and her four children, meanwhile, continue to sleep on mats on the dirt floor of her one-room adobe structure. US remittances have enabled many mothers to stop the daily labour of making tortillas and selling them in local markets. Luisa, however, continues to work over the hot *comal,* the roughness of her hands testimony to the toil of tortilla-making where the skin of neighbours' hands has been smoothed over with the flow of 'global' capital. Luisa's children are teased at school because they cannot afford uniforms or supplies. They often stay at home to work with their mother, and her oldest son leaves home to do whatever odd jobs he can find in the village: tending animals and collecting firewood.

Figure 14.3 Oaxaca is one of Mexico's poorest states. Many of its residents have migrated to the US in search of jobs.

The daily lived experiences of transnational residents of Oaxaca demonstrate that globalization is a phenomenon with social, cultural, economic and political dimensions. This village boasts various businesses that enhance communications and travel between places. Relationships, too, transpire transnationally. Village festivals to celebrate the patron saint offer a time when migrants return home, investing remittances in new outfits and construction projects and extravagant parties. Of course these celebrations are videotaped and sent to those in Poughkeepsie who were not able to make the journey (Mountz 1995).

Although located amid this new influx of income, Luisa's family experiences economic hardship and corresponding social marginalization in the village. Whereas many of the neighbours participate in conspicuous consumption using substantial remittances from family members in the US, Luisa and her children do not have the material resources to contribute to the increasingly extravagant village festivals. Luisa receives cash remittances from her husband only two or three times a year, and the amounts have proven inadequate to support the family. Luisa suspects that her husband's earnings now support a new family in New York while she struggles to feed, clothe and maintain the health of her children (Mountz and Wright 1996).

Luisa herself would like to participate in the global economy by traveling to Poughkeepsie. But as the woman head-of-household, she has four children to care for and is unable to leave them behind so that she can work in the United States like her husband. Her location and material circumstances suggest that not everyone benefits from globalization and the global economy. In the intimacy of the daily struggle with poverty, households without remittances are unable to contribute to or benefit from the collective structures that channel and manage new resources in the village. Whereas time-space compression may have brought Oaxaca and Poughkeepsie closer together (Mountz and Wright 1996), not all residents of this expansive transnational space will benefit equally.

KEY POINTS

- Globalization involves the intensification of flows (of people, goods, ideas, capital) across local, national and regional boundaries.
- Globalization involves what geographer David Harvey (1990) calls 'time-space compression'. Modern technologies (the Internet, fax, aeroplane, etc.) have decreased the time it takes to connect people in different parts of the globe. Distance is less of an obstacle so space has become compressed.
- There are three debates in the globalization literature.

1. Some scholars believe globalization has destroyed the nation-state. Others believe the nation-state continues to exert influence, albeit in different ways.
2. Some scholars believe globalization is leading to cultural homogenization or McDonaldization across the globe. Others believe that local cultures are hybridizing – adopting elements that fit or blend, and rejecting those that do not.
3. Feminists argue that studies of globalization have tended to privilege the global over the local, and as a result have overlooked how global processes are embodied differently across social categories such as gender and race.

FURTHER READING

Dicken, P. (2003) *Global Shift*. New York and London: Guilford Press.

Held, D., McGrew, A., Goldblatt, D. & Perration, J. (1999) *Global Transformations: Politics, Economics, and Culture*. Stanford, CA: Stanford University Press.

Massey, D. (1993) 'Power-geometry and a progressive sense of place', in J. Bird, B. Curtis, T. Putnam, G. Robertson & L. Tickner (eds), *Mapping the Futures: Local Cultures, Global Change*. New York: Routledge. pp. 59–69.

Nagar, R., Lawson, V., McDowell, L. & Hanson, S. (2002) 'Locating globalization: feminist (re)reading of the subjects and spaces of globalization', *Economic Geography*, 78(3): 257–284.

15 MIGRATION

Alison Mountz

Definition: A Concept Traditionally Defined through Dichotomy

Migration is defined as movement from one place to another. All sorts of species, human and non-human, migrate. Human geographers and other social scientists have focused primarily (though not exclusively) on human migration, the movement of individuals and populations from one place to another. Indeed, migration is an important part of the human condition. As much as people long for a sense of home (Dowling and Blunt 2006) and place (Tuan 1977), they also move for a wide array of reasons.

While migration is widely discussed and politicized in public discourse as a pressing *contemporary* issue, it is certainly not new and must be placed in its historical context. People have always migrated. Long before nation-states existed, indigenous tribal groups moved in search of subsistence, whether hunting, grazing, or migrating seasonally in cycles when the agricultural lands in one site were no longer productive. Archaeologists have pieced together historical narratives of great human migrations across continents (see Castles and Miller 1998). Although migrating peoples may have come into contact with others with whom they traded wares and food and at times fought over territory, human migration itself only became categorized, politicized, qualified, quantified, studied and controlled with the growth of the contemporary nation-state. Throughout the nineteenth and twentieth centuries, human mobility evolved and intensified dramatically with the development of technologies and industries designed to transport people, on the one hand, and to regulate human migration, on the other. This growth continues today and its urgency – brought on by the 'immediacy' of globalization – puts pressure on debates about migration and citizenship.

Nation-states have an interest in regulating their citizenry, a group of people defined by their attachment to a national territory, bound by national borders, and issued legal identities that are documented and catalogued with identity documents such as passports. The regulation of migration is primarily the domain of nation-states. John Torpey (2000: 3) goes so far as to argue that 'the emergence of passport and related controls on movement is an essential aspect of the "state-ness" of states'. Many states manage migration with immigration and refugee policies and border control.

Fewer states regulate migration within sovereign territory. China, for example, has attempted to regulate migration among its sizeable national population by requiring people who want to move to receive government approval to do so (Fan 2004). In Palestine, a territory that is not recognized as a state by many other states, internal migration is extraordinarily difficult. Migrants, even those only traveling small distances, must pass through heavily militarized checkpoints.

Still other states seek to regulate migration beyond sovereign territory for a variety of reasons. Through colonial projects past and present, states use force to occupy distant regions and impose borders (Gregory 2004). States also intervene, sometimes unilaterally, though more often through multilateral arrangements to impose peace and to protect refugees displaced from regions in conflict (Hyndman 2000). In other instances, states embark on imperial projects that relate less to conflict or humanitarian assistance and more to their own interests in stopping people from entering their own territory (Nevins 2002). These projects have increasingly been referred to as the 'securitization' of migration: the exercise of control over human mobility abroad, primarily in the interests of national security at home.

If exercises in categorization and control are central to the project of defining migration, it is important to begin by naming their terms. Migrations can be divided spatially as internal (within sovereign territory) or international (across international borders). Most contemporary debates and research studies on human migration involve migration across international borders. Migrations are also catalogued temporally, designated as either permanent or cyclical. A person who moves from India to Washington, DC, to become a citizen of the United States, for example, would be considered a long-term migrant. 'Snowbirds' (McHugh and Mings 1996), elderly North Americans who travel south to live in trailer parks in summer, or seasonal agricultural workers who follow crop harvests, would be considered cyclical migrants.

Human migrations are also often categorized according to their root causes. Voluntary migrants might move to work, study, or join family members. Involuntary migration, on the other hand, involves the displacement of a population forced from their homes due to conflict or persecution. The creation of the Israeli state in 1948, for example, displaced Palestinians who had lived on the land previously. During the civil war from 1979 to 1992 in El Salvador, political persecution and the violation of human rights prompted approximately one million people to flee their homes. Some resettled regionally in Guatemala and Mexico, while others moved to Europe and North America.

Migration is often discussed and regulated using a range of binaries. Nation-states often facilitate legal migration so that people come to work, filling niches in the labour market, whereas other migrations are deemed illegal, also referred to as unauthorized, undocumented, or unmarked. This distinction between lawful and unlawful movement demonstrates the primacy of states in the regulation of human mobility. States set laws and determine who can access citizenship or temporary legal status.

Policies applied to people on the move divide them in other ways too. Some people will be considered temporary migrants and granted temporary status. They probably travelled to study or work for a limited time, or were granted temporary status to reside in a place while an application for asylum was adjudicated. Others will be considered more permanent immigrants who have come to stay. They may have come as highly skilled professionals, as family members of immigrants with citizenship or permanent residency, or as refugees resettled from abroad.

States seek to manage human migration by developing policies that correspond with many of the categories suggested above. These categories signal exercises in identification that are often contested by individuals and groups and cause a great deal of debate, dissension and subversion, often manifested in struggles over border crossings (see Figure 15.1).

Evolution and Debate: 'Migration wars'

Migration is one of the most dramatic of human experiences, so discussions about it are often hotly contested. Jennifer Hyndman (2005) has called these debates 'migration wars'. By 'migration war', she means fights over who to let in and who to keep out. These debates are politicized in public discourse and manifest in heavy investments in border

Figure 15.1 The border crossing between Tijuana, Mexico, and San Diego, California in the US. In this photo cars are lined up on the Tijuana side of the border, waiting to enter the US.

enforcement in order to patrol and police the boundaries of sovereign territory (Figure 15.2).

As a result, migration is a key area of research within political geography and gives rise to much debate and public discourse in contemporary societies. Migration finds its way into election campaigns by politicians eager to advance political or economic agendas (Bigo 2002). The terms of debate are frequently structured by the dichtomies named above. As human smuggling became more widely debated and discussed in the 1990s in the European Union, Australia and North America, for example, federal governments discussed the desire to 'manage' migration. Much nationalist rhetoric involved a desire to 'crack down' on illicit forms of migration in order to enable a more orderly, managed, flow of immigrants who were accessing legal channels.

Some immigration debates relate to demographics and have long been research topics of population geographers who study mortality, conception and fertility (Bailey 2005: ix). Many western countries,

Figure 15.2 Electric fence at Baka-El-Garbia separating Israel from Palestine. The border between Israel and Palestine is one of the most fortified borders in the world.

particularly member states of the European Union, face declining rates of 'natural' population growth. Without immigration from other countries, not only will the population decline dramatically in Italy, Spain and Canada, but the national economy will suffer as there will not be enough people to fill jobs in the labour market (Ley and Hiebert 2001). Local and federal governments are invested, therefore, in recruiting workers into specific niches in the labour market, nurses, high tech workers and domestic workers among them. They design specific immigration programmes and recruit workers from other countries. They also sponsor *internal* migration programmes, rewarding citizens and immigrants who move to remote locations where workers are needed; in the remote north country of Canada, for example. There are not enough skilled migrants to administer government programmes, and there is a desire to populate and govern the area to service and support populations who already live there and to protect sovereign interests in the sparsely populated but resource-rich and strategically located Arctic region.

Those who immigrate as recruits of worker programmes or to work or invest are called voluntary migrants and economic migrants. They are motivated to move for economic reasons and generally do so of their own volition. They might feel that they have no option but to move to make a living, but they have not been forced to leave their home country by violent means.

Those forced to leave home are called involuntary migrants. They are displaced from home, often by armed civil conflict, persecution on the basis of their identities or – increasingly – environmental disaster. Those seeking protection are political migrants or refugees. They are designated 'political' because they were forced to move as a result of persecution. Speaking generally, there are two kinds of displaced person: refugees and asylum seekers (also called refugee claimants in some countries). Refugees are persons who flee their home country and often find temporary refuge in refugee camps. Their best chance of survival is through aid and local work in the short term, and resettlement by a 'safe third country' in the long term. Refugee-receiving countries with managed resettlement programmes work alongside the United Nations High Commissioner for Refugees to interview and resettle a very select and small percentage of persons housed temporarily in refugee camps, primarily in Africa and Southeast Asia (see Figure 15.3).

Asylum seekers or refugee claimants sometimes flee their homes in search of protection, and will travel to sovereign territories of countries with asylum programmes and request protection upon arrival by making a claim. They may be granted temporary status until their case is heard. Some are detained (particularly in Australia, the United States and the United Kingdom). A significant percentage are declined, and are either returned home or commonly stay in the host country illegally, undocumented by the state.

Recent trends involve efforts by states to increase border control and reduce the numbers of asylum claimants who reach their shores. Many do this with 'externalization', attempting to process and detain asylum seekers offshore where they have fewer rights to access systems of protection (Mountz, 2009).

Refugee policies are highly dynamic in a number of ways. While the general definition of a refugee was set by the UN *Convention Relating to the Status of Refugees* in 1951 and its 1967 Protocol, it is up to

Figure 15.3 Refugee Camp in Northern Sudan

signatory countries to decide how to interpret, manage and adjudicate asylum-seeking populations. Some countries are more liberal in their definitions than others. Canada, for example, adjudicates refugee claims of gender-based persecution, awarding refugee status to women who face forced sterilization and genital mutilation in their home countries. Canada also now grants refugee status to persons with non-normative sexual identities who have fled countries where they can prove – like all refugee claimants – a well-founded fear of persecution upon return on the basis of gender or sexual identity.

Some of the most fierce and public wars are fought over undocumented or illegal migration. While countries seem to agree that they need workers, they are often at odds about how best to pursue this goal. Whereas temporary guest worker programmes gained popularity in European Union countries in the 1990s, similar programmes were not established in North America. The result has been a significant growth, especially in the US, in undocumented migrants (estimated at 12 million).

Since the late 1990s, and especially following the terrorist attacks in 2001, the regulation of migration has become increasingly securitized. Those in search of protection after being displaced for political reasons often find themselves caught up in security agendas. The countries with the largest managed immigration and refugee programmes are also those with some of the most sophisticated border enforcement programmes. While they are likely to settle a certain number of refugees from camps abroad, they are less accepting of what they call 'spontaneous arrivals': those who arrive on sovereign territory of their own accord and then make an asylum or refugee claim for protection.

Another escalating set of 'migration wars' involves human smuggling and human trafficking. These are opportunistic industries that often arise symbiotically in relation to border enforcement; stronger enforcement measures to keep people out prompt a greater demand for agents of human smuggling (Chin 1999). Human smugglers take different forms in different countries and cultures: so-called 'Snakeheads' move Chinese migrants to Europe and North America by boat and plane while 'Coyotes' move South and Central American migrants across northern land borders into the United States and Canada.

As these industries have grown in recent decades, so too have efforts by national and supranational authorities to regulate migration in new ways. Domestic and international laws and protocols are seeking to protect the human rights of those trafficked while setting increasingly tough penalties for smugglers who profit from human migration (Kyle and Koslowski 2001).

Meanwhile, social scientists are also finding new ways to understand the dynamic field of human migration. Throughout the 1990s, interest grew in the area of transnational migration: populations on the move in more cyclical ways and living lives with daily transnational dimensions characterized by remittances to support family and development projects at home and daily decisions that connect and contextualize life's decisions in relation to multiple locales across borders.

Case Study: Hmong Refugees and Categories of Migration

Closer examination of people's daily experiences of migration and displacement exemplifies these categories of migration, while at the same time exposing problems with some of the categories that inform immigration and refugee policies. Geographer Ines Miyares's (1997) research on the Hmong refugee population in the US demonstrates both the usefulness and blurriness of many of the migration categories outlined thus far. The Hmong are a Southeast Asian semi-nomadic population historically found in large numbers in China and Laos. Prior to their political displacement Hmong migration patterns related to slash and burn agriculture. The Hmong way of life was disrupted as a result of their involvement in the Vietnam War; many Hmong assisted US military operatives, believing that they would be offered a homeland in return. Instead, they suffered attacks by the Pathet Lao Communist government. Many fled to refugee camps in Laos and were later resettled in the US, starting in 1975 (Miyares 1997: 216).

The largest Hmong population outside China lives in Fresno, California and numbers approximately 35,000. As Miyares notes, Hmong migration patterns demonstrate that the group does not fit easily into one migration category. As refugees resettled from abroad by the US government, they are international migrants who were displaced from their homes involuntarily. After being resettled across the United States, however, they became internal migrants, embarking on 'secondary migrations', moving west away from the midwestern American cities where they were resettled to voluntarily form enclaves in Californian cities. Fresno, located in the San Joaquin Valley in California, is home to a number of Hmong refugees.

Miyares conducted in-depth interviews and participant observation among 120 Hmong university students and their families in California's San Joaquin Valley. She looked especially at the process of acculturation through which refugees adapted and became socialized to life in their 'host' country. She found that in spite of repeated migrations that spanned many years and crossed multiple national borders, some Hmong refugees continued certain cultural practices from the past and synthesized these with practices adapted to the refugee camp and then resettlement experience. In particular, Hmong resettled in the United States reproduced aspects of their semi-nomadic lifestyle with 'simple and portable furniture, typically woven mats for sleeping and on which food was prepared' (1997: 216). Miyares identified these practices by studying 'shared perceptions of the concepts of home, space, and place' (1997: 214). In refugee camps where space was limited, Hmong families shifted from slash and burn farming to the intensive farming of small plots of vegetables and then traded their surplus at local Thai markets (1997: 217). Once in California, while they were not usually allowed to cultivate the land outside apartments, many kept small window boxes or other small plots available to grow fruits, vegetables and spices. They participated economically in small-scale farming and produce sales in flea markets – the Fresno market being the largest one – where traditional forms of bartering and other social relationships continued (1997: 219).

The Hmong reproduced the ability to pack up home and move quickly at short notice and in relation to clan networks among communities. According to Miyares, most belongings unnecessary for daily living remained packed in boxes that were stacked along walls (1997: 220). The Hmong came together for Hmong New Year, which remained an important holiday where the community overcame distance to celebrate survival, arrange marriages and exchange gifts, among other traditions.

In short, Hmong migration (along with most international migrations) proved political and economic, international and internal, temporary and permanent. The political displacements that led to refugee resettlement also resulted in hybrid cultural forms that joined cultural practices from the past with practices that enabled the population to adapt to resettlement.

Future Research

One of the reasons that migration is such an important area of study is that it traverses many boundaries, connecting in its wake so many other social, cultural, economic and political phenomena. For many decades, Mexican migrants moved to the United States to work. More recently, American companies shifted their manufacturing sites to Mexico in search of less environmental and economic regulation. As globalization continues and the world economy shifts, China and India grow more powerful economically. How will human migration change as a result? Where will workers travel in the future? How will nation-states continue to both adapt to and intervene in these global landscapes of migration? These are among the questions being asked by political geographers.

Of particular interest to those who study refugees is a group increasingly referred to as 'environmental refugees'. As global warming continues, climatologists predict that coastal parts of Bangladesh and other low-lying countries and regions will be submerged in water, forcing significant populations to move. Still other populations will need to move in order to seek new livelihoods as climate change affects their economy and ability to grow food. In early 2008, the United Nations released a report warning that some 25 to 200 million persons could be displaced for environmental reasons by 2050. While there is increasing concern about these potential displacements, the United Nations High Commissioner for Refugees has been reluctant to categorize this group as refugees because they do not fit the traditional definition of a refugee, defined as someone persecuted for political reasons.

Geographers will continue to play a key role in understanding human migration, precisely because of the multiple methods they use to study human phenomena, the interdisciplinary nature of geographic inquiry, and the unique connection between the natural, cultural and political worlds that the discipline examines (see Graham 1999; McHugh 2000).

KEY POINTS

- Although migrating peoples have come into contact with one another for centuries, human migration only became categorized, politicized, quantified and studied with the growth of the contemporary nation-state.
- Political geographers study so-called migration wars – fights (usually by nation-states) over who to let in and who to keep out.
- As globalization proceeds, countries have invested heavily in border enforcement, increasing patrols along borders, building fences, and using infrared technology to keep people from crossing borders illegally.
- When borders become more difficult to cross, human smuggling and trafficking tends to increaase.

FURTHER READING

Bailey, A. (2005) *Making Population Geography*. London: Hodder Arnold.

Castles, S. and Miller, M. (1998) *The Age of Migration: International Population Movements in the Modern World*. New York and London: Guilford Press.

Hyndman, J. (2000) *Managing Displacement*. Minneapolis: University of Minnesota Press.

Part IV Bounding Space

INTRODUCTION

Carolyn Gallaher

The three concepts covered in this part discuss the ways that political geographers conceptualize the relationships *between* places. In particular, we look at two relationships – connectivity and division.

The first chapter of this section, on **scale**, captures both concerns at once. When geographers think of space, they tend to envision it as different scales: the local, the regional and the global. However, scale is a social construction inasmuch as one cannot find scale, or its boundaries, on the landscape. It is a heuristic device for thinking about how places can be grouped together, and, by contrast, how they can be delineated. The concept of scale allows us to generalize about places and the relationships between them.

The second chapter in this part, on the **border**, focuses on the division of space into separate places. Unlike scale, which is largely invisible on the landscape, the border is readily apparent on the landscape. Borders are often demarcated by walls, fences and military agents. Borders can be difficult to cross, especially for people moving from a poorer place to a wealthier one, from a violent one to a peaceful one. Geographers examine both the physicality and the symbolic meaning of borders. They study how borders are enforced as well as what those standards of enforcement mean for those on the inside, and those outside them.

The final concept chapter in this part is **regionalism**. In political geography the term 'region' is used to describe a number of places which share a common feature. At the macro level, a region includes several states which share common features. The term Latin America, for example, is used to describe the part of the Americas which was colonized by Spain and Portugal. Regions can also refer to subnational spaces, micro-level areas which normally contain a shared language/history/culture, etc.

In this part of the book examples are drawn from a variety of places across the globe. The US elections are discussed. Contentious borders in Europe, the Americas and Africa are detailed. And the region of Kurdistan, sometimes known as the largest nation without a state, is analysed.

16 SCALE

Carl T. Dahlman

Definition: Describing Spatial Relationships

The geographical meaning of scale is commonly understood as the extent or size of a phenomenon. Major events such as World War II or the 2004 Indian Ocean tsunami are termed large-scale because of their effect over a large area of the earth's surface. The terrorist attacks of 11 September also may be considered large-scale because of their extensive indirect effects – both psychological and geopolitical – despite having affected a very small physical area. In contrast, an electoral campaign for the city council of Helsinki is considered a small-scale event because it affects a relatively small population area, even if the issues and outcomes are of greater daily importance to those voters than either the tsunami or 9/11. Thus, we often conceive of a large-scale phenomenon, such as globalization, global climate change or the 'war on terror', as being of great importance for everyone on earth even though such issues may rank much lower on any one individual's priority list (Figure 16.1). This disjuncture between geographical scale and political importance is key to understanding why geographers have tried to invest the term with greater conceptual rigour.

Geographers, therefore, use scale in several more specific ways, each representing some basic spatial relationship. The first draws from the technical use of the term in cartography and allied techniques in which scale is a mathematical expression of the relationship between the map and the real world. This is often expressed as a ratio of map units to real units: for example one inch on the map represents one million inches on the ground. Although scale in this sense denotes a proportion of measurement between the map, as a representation of the world, and the world itself, scale also implies a simplified representation of the world. The more surface area included on a map, the greater the need to select and generalize features.

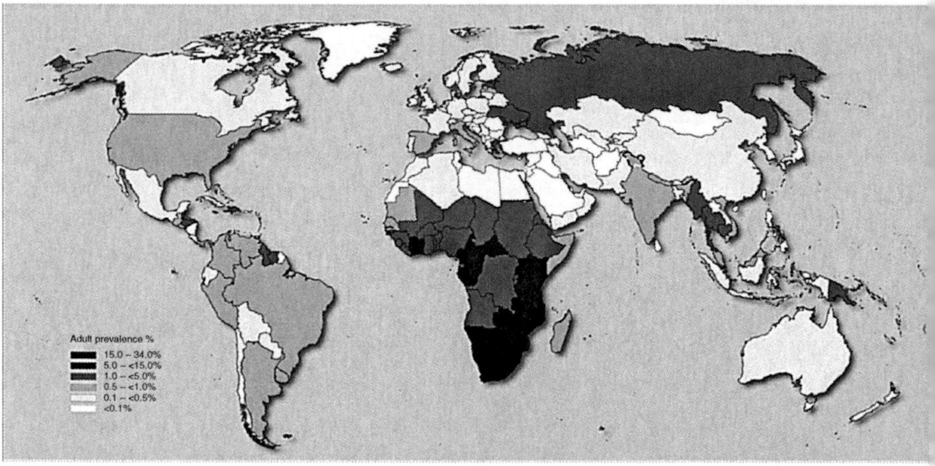

Figure 16.1 The prevalence of HIV infection is a worldwide phenomenon while also being distributed unevenly among the world's population UNAIDS, 2005.

The second meaning of scale relates to methodological concerns, that is, the unit of analysis and level of resolution required for a geographical study. This meaning relates closely to the cartographic meaning of scale, but here it is a logical relationship between a research question and the level of observation. It is not possible, for example, to answer questions about individual voter attitudes by looking at summary election data for a whole country (Quattrochi and Goodchild, 1997). Similarly, one cannot make generalizations about international migration patterns by studying just one migrant.

A third, and more prevalent, meaning of scale is as a 'vertical' series of nested levels, local, national, regional and global, that provides a convenient way of thinking about relationships between humans and institutional actors across different spatial extents. Some political geographers have viewed these nested levels as mirroring top-down or hierarchical power arrangements. Peter Taylor has argued that the global is the most important scale because it is the operational level of the world economy (Taylor 1982). Susan Roberts, in contrast, argues that our tendency to view the local as the necessary outcome or product of the politics 'above it' is not only incorrect but stems from a gender-biased interpretive framework that animates the global as the 'masculine' scale of power and action, rendering the local as a feminized, passive product of global interactions (Roberts 2004). Indeed, bottom-up scalar relationships are also important, as on election day when local and state ballot issues affect who goes out to the polls and how they vote on national races that also appear on the ballot.

Distinct from but closely related to the idea of scale as nested levels, a fourth sense of scale demands that we focus on the processes that produce the relationships between places. This idea is most often given form in the claim that relationships are scalar; in other words the operational aspects of human interactions in space create those scales (Marston 2000: 220). Economic relations, treaties and cultural norms are examples of social processes that manifest themselves in spatial forms, especially within existing institutional settings. The General Agreement on Tariffs and Trade (1947), for example, eventually led to the World Trade Organization, a forum at which 151 member states hammer out international trade rules, creating a near-global scale of inter-governmental interaction. These processes may also challenge existing institutional arrangements, however, and may even shift the scalar level at which they operate. Thus although trade has been international for hundreds of years, the role of the state in its promotion and regulation is changing under the WTO. Technological shifts in transport and communications and the growth of transnational corporations have combined with neoliberal agendas to lessen the role of the state as a major actor in economic relations. (For more information on neoliberalism see Chapter 13.) Although by no means a *fait accompli*, the significance of spatial shifts in the organization of the global economy requires that we pay attention to its changing operational scale.

Evolution and Debate: From Empty Stage to Active Space

Scale has proven to be a useful term in geography because it recognizes that although the world is made up of almost infinite levels of detail, there is a need to generalize or simplify these details so that we may in turn recognize important spatial trends, relations and dynamics in human affairs. The careful attention given to what we mean when we invoke scale is a recent phenomenon, however, and many geographers remain just as likely to invoke scale as a taken for granted and static vertical organization of the world (Taylor 1982). Some authors have rightly objected to taking scale for granted, arguing that it is incorrect to think of scale as an unchanging natural stage on which society acts. There is no a priori state or regional scale, for example; they are human inventions. Instead, these authors argue that scale should be understood as a way of framing the world, an interpretive process by which humans make sense of events and relations that extend well beyond the realm of individual experience yet impinge on their lives.[1] Scale, according to Delaney and Leitner, is 'not simply an external fact awaiting discovery but a way of framing conceptions of reality' (1997: 94–95). Furthermore, this sense of scale means that it is also a form of politics itself, as when competing interests disagree about the right 'level' for making decisions within a polity (Agnew 1997). The US Constitution, for example, is based on the idea that certain decisions are assigned to either the federal or state government. The ability to define scale is really an ability to define the boundaries and relations that obtain on a given issue (Smith 1992). Consequently, the outcomes of decisions about how to allocate public and private goods have material effects that in turn embed scale in society.

Some political geographers have further argued that scale is only meaningful as a human construct, continuously interpreted and reasserted in the process of thinking about and acting in the world. This view accepts scale as primarily a category of meaning that comes from social and political contests, rather than a necessary reality. Approaching scale as an interpretive framework of political convenience, however, leaves out the ways in which ordinary lives are reproducing scale on a daily basis, through spatial choices of where to live, work, play. In other words, it is not sufficiently useful to think of the production of scale by powerful institutions as a one-way process, shaping the lives of the masses with predictable effect. We should instead, argues Sallie Marston (2000), focus on the ways in which households and individuals make decisions that also contribute to the production of scale. The aggregate effect of these decentralized practices might thus be formative agents in making and modifying scalar frameworks once thought to operate from the 'centers of power'. We might also, then, be able to identify the sites and scale of resistance that exist in the prevailing social, political and economic order.

Another, more recent development in discussions of scale, at least within Anglo-American human geography, is a more contentious suggestion that we 'eliminate scale as a concept in human geography' because of the conceptual problems it presents (Marston et al. 2005). In addition to 'flattening' our view of society, a small cadre of authors argues that we should eliminate horizontality, embracing only a network approach – the empirical connections between people, institutions and places – for studying human geography. They argue that by focusing only on the complexities of social realities in a particular place, it is possible to avoid oversimplifying the social contents of a site. This goes beyond conceptual debates of scale's abstract vertical levels and its horizontal extent, to jettison the idea altogether (Taylor 1982). The call to evacuate human geography of scale is of course provocative and yet for all its postmodern pedigree, this approach does more to express an anxiety to 'get space right' than to improve our understanding of how humans use scale in their lives.

As this term seems to belong to geography, there is a degree of disciplinary self-interest in trying to infuse the concept with greater rigour than it can perhaps bear. It may be better to consider the practical value of leaving scale as a necessarily contested abstraction. There is, as well, a risk in abandoning scale – that human geography will be unable to address processes that are conceived, implemented and commonly understood according to scales, however problematic such conceptions can be. With this in mind, David Prytherch uses the example of Wal-Mart's enormous material flows and effects as a bundle of economic and urban processes that simply cannot be understood without resorting to geography's better developed concept of scale (Prytherch 2007). Specifically, he argues that it may be more useful to think of scale not as size or extent but rather as *proportion*, always in relation to the body's space, on one level, and the finite expanse of global space, on another. In this way, we are reminded that terminology in human geography is meaningful only as it relates to human existence in the closed system of terrestrial space.

Thus, scale is at once a simple idea and a slippery term subject to debate among human geographers. The reason for this is that scale attempts to provide a framework for describing the relationships between places but these relationships are never independent of geographic processes envisioned and enacted by humans. In other words, scale conveys a sense of mathematical precision, but human relations are far too messy and uncontained for the term to line up with reality. Yet even if scale oversimplifies society it nevertheless provides an important means of generalizing the complexity of human relations that produce the places and spaces of interest to geographers. Although this conundrum has sparked much discussion, the necessary convenience of one of geography's 'key terms' means that scale remains a useful idea, if without consensus on how (Leitner and Miller 2007).

Case Study: American Elections

Elections provide numerous illustrations of how scale operates as an embedded aspect of social reality. Constitutionally, states are internally ordered by granting superior and inferior powers to multiple representational and administrative levels. These powers include the organization of representational government, the jurisdictions of law, and the mechanisms of allocation and enforcement. In the United States, constitutional federalism assigns powers to two primary scales, the national and state. The constitutions of the 50 states and US territories further provide for various local governments (e.g. city councils, county governments), some of which may have overlapping areas but separate responsibilities. Enfranchised citizens are thus able to participate in the election of representative bodies and vote on referenda on at least three scales. Elections and the governing authority that results from them are a good example of how social processes actively produce scale and give it meaning. These scales, however, are not independent of one another. For example, even though the constitution reserves powers to the states, their laws must not conflict with US federal law.

Elections associated with different scales of government, although technically unrelated, are to varying degrees interdependent. Because voting in different races and on various issues happens at the same time and in the same voting booth, individual voters are prone to a variety of inter-scalar effects, meaning that their separate selections are somehow interrelated. Popular presidential candidates, for example, often exert a 'coat-tail effect' on other candidates 'down ballot' within their political party but running in other elections, even if their campaign positions diverge. For example, a popular Republican candidate running for President attracts voters to the poll who would not normally come out to vote and who then vote for other Republicans appearing on the ballot in state or local races. More importantly, while the conduct of elections is subject to federal and state law, the act of casting ballots falls to locally constituted boards of election.

The 2000 general election highlights some of these issues in the complex legal battle of counting ballots in the state of Florida. At the end of polling on election day, the Republican candidate, George W. Bush, and the Democratic one, Al Gore, were effectively tied at the national level and were waiting for the outcome of the election in Florida, where they were also separated by a very small number of votes. Under the US Electoral College system, the candidate with the largest number of votes in a state wins the whole state. When a small number of Florida precincts ran into trouble with their voting procedures – which involved enough votes to decide the outcome in Florida – the outcome of the entire election was put on hold. In particular, a number of voting irregularities related to Florida's local election board procedures raised questions as to whom the voters had actually chosen for President. At issue were unreliable ballot technologies used in some precincts where voters had to line up a blank punch card with a fixed list of candidate names and then punch a hole in the card next to the name of their preferred candidate (Figure 16.2). Many of the cards didn't line up correctly and other voters failed to completely perforate the card, leading to a dispute over how to count votes in those precincts.

The remedies provided for under local and state law were soon exhausted without resolution and the ballot dispute went to court. Because state law was involved, the legal cases moved through the Florida courts before being heard by the US Supreme Court. The Supreme Court's decision to stop the recount in the disputed Florida localities decided the race for Bush. This case raised important questions about the scalar relationships of election law, especially related to civil rights and voting procedures, not least because the Florida official in charge of elections worked for the Bush campaign. It also triggered a debate about the need for national standards in local voting machines, which remains fraught today. What this crisis laid bare was the scalar relationships in operation between political units engaged in the election process, as well as in the legal jurisdictions created under US federalism. The débâcle led to calls for the direct popular election of US presidents, effectively jumping the state scale. This would require, however, some mechanism to get around the election role assigned to states in the Electoral College under the US Constitution.

The reality of election day voting behaviour has also shown how candidates try to 'game the system', taking advantage of the multi-scalar moment in the voting booth. In the 2004 general elections, US voters were presented with a choice for President and Vice-President, US representatives and, in some states, US senators, as well as state and local representatives. In some locales, additional measures were added to

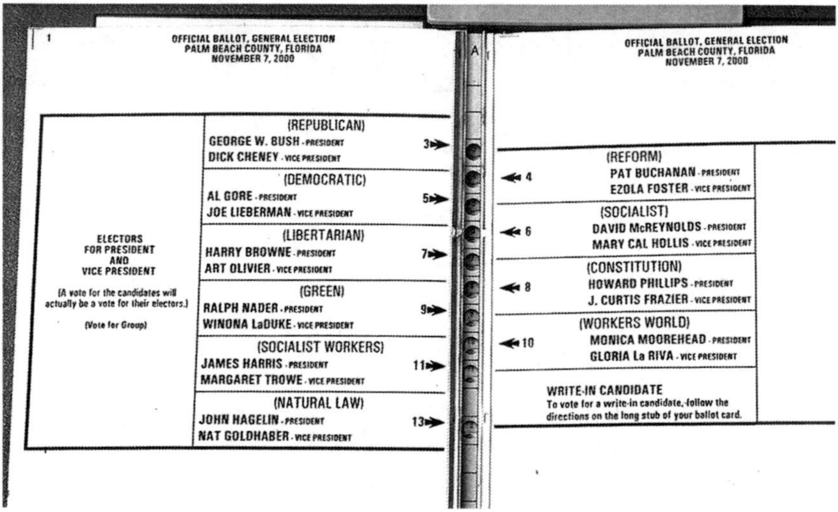

Figure 16.2 The "butterfly ballot" used in a disputed Florida precinct. Voters were to slide a card behind the ballot and, using a punch-pen, indicate their selection. Since Gore was the second choice down the left side but the third hole down the centre, Gore supporters may have erroneously punched the second hole, casting their vote for Buchanan.

the ballot that allowed voters to directly decide issues of local or state interest. Among these were such issues as school tax levies and bans on public smoking. Some states also included ballot measures dealing with same-sex marriage, an enormously controversial issue that was also the subject of extensive media attention. During the 2004 campaign season there was a much publicized and scrutinized decision to allow same-sex marriage in Massachusetts, as well as a decision by the Mayor of San Francisco to issue marriage licences in spite of Californian law. In response, politicians and groups opposed to same-sex marriage or civil unions organized state-level 'defence of marriage' referenda to appear on ballots in 11 states. These measures either prevented the states from permitting same-sex unions or else limited marriage to opposite-sex couples. The media attention given to gay marriage helped energize conservative voters who had begun to doubt Bush's presidency and who might not have shown up on election day were it not for a distraction issue like gay marriage. Even where anti-gay-marriage measures were not on the ballot, the pre-election focus on this part of the

US 'culture war' helped boost conservative turnout, thereby aiding in the re-election of President Bush.

KEY POINTS

- Geographers and others use the term scale in several different ways, although each describes a geographical relationship.
- Scale as it relates to political geography describes how we frame the world and function in it according to the relationship between 'levels' of political units, jurisdiction and identities.
- Scale is also understood to be dynamic, created or at least shaped by political contests over control of space. It also helps relate political geography to economic, social, transnational and natural phenomena.

NOTE

1 See especially the erudite review of social constructionist approaches to scale in Marston (2000: 221–233).

FURTHER READING

Delaney, D. and Leitner, H. (1997) 'The political construction of scale', *Political Geography*, 16(2): 93–97.

Herb, G. H. and Kaplan, D. H. (1999) *Nested Identities: Nationalism, Territory, and Scale*. Lanham, MD: Rowman and Littlefield.

Marston, S. A. (2000) 'The social construction of scale', *Progress in Human Geography*', 24(2): 219–242.

Roberts, S. M. (2004) 'Gendered globalization', in L. Staeheli, E. Kofman, and L. Peake (eds), *Mapping Women, Making Politics: Feminist Perspective on Political Geography*. New York: Routledge. pp. 127–140.

17 BORDER

Alison Mountz

Definition: Rigidity and Fluidity

To some, a border is a line drawn on a map, materialized on the ground in a fence, hedgerow, sign or checkpoint. It does the work of distinguishing between two separate entities on either side. For political geographers, the border is often international; it separates the national territories of two nation-states.

But in reality, a border takes many forms and serves all kinds of functions. The border operates in as many ways and places as one's geographical imagination can fathom. The border certainly *is* a line that delineates here and there, separating an 'us' from a 'them', one place from another. It is physical, tangible, material. For evidence, one need only witness the militarized construction of fences and walls along the border separating the United States from Mexico, North from South Korea (see Figure 17.1), or between Israel and Palestine today (Figure 17.2). As these examples illustrate, borders often have an infrastructure to regulate crossings, in many cases exacerbating material inequalities between persons on one side or the other. The walls that surround and separate a gated community from its neighbours, for example, often emphasize the greater wealth of those who live behind the gates (see Figure 17.3). Likewise, an urban university campus with walls around its perimeter demarcates the divide between students and their neighbours.

Various industries have arisen to both maintain and traverse borders. Border enforcement, for example, is undertaken usually by government authorities or private security companies that patrol borders and use sophisticated technology, such as infrared sensors, to catch those making illegal crossings. Other groups work to subvert border enforcement, smuggling humans, drugs and money across borders. In fact, the business of smuggling people across borders now rivals the lucrative business of transnational drug smuggling (Kyle and Koslowski, 2001). In short, border crossings are desirable, and profits are high in the industries that capitalize on this desire.

Figure 17.1 Fence topped with barbed wire in the demilitarized zone between North and South Korea

Figure 17.2 Concrete wall separating Israel from Palestine Bakka-El-Garbia, Israel

Figure 17.3 The front entrance to a gated community in the US

The border signals division and the containment of people, ways of life, economic goods and systems to regulate trade. Borders demarcate not only the contours of the nation-state, but also a highly visible location where the state is able to flex its might to protect, police, regulate; where states express sovereignty, often in dramatic ways. The growth of the machinery to enforce it is often referred to as *militarization*. Political geographers have studied the militarization of borders between nation-states, particularly as sites where conflicts over territory transpire.

Of course borders are also drawn so that they can be crossed. They are not solid lines, but porous thresholds. The border is a meeting place of conjoinment where persons and ideas come together; where people and places meet, abut, mingle and neighbour. At the same time that the United States' government increases security along its southern border with Mexico, for example, thousands of workers daily cross the border to do jobs on the other side. Many work in the booming service industry on the US side of the border, while others work in the *maquiladoras*, foreign-owened factories near the Mexico-US border where corporations enjoy reduced tariffs and duties and cheap labour.

The border slices through areas known as 'border regions', and many studies have been done of border regions that ask, for example, how they are governed and what kinds of crossings, cultural hybridities and transnational businesses exist there. Sometimes nations and communities span borders, whether first nations groups that predate national borders or international travelers holding dual citizenship who spend their time betwixt and between places (e.g. Nevins 2002; Sparke 2006, Sparke et al. 2004).

Evolution and Debate: Bound for Controversy

Borders are contradictory, paradoxical sites, and the debates about them are no less confusing. Joe Nevins (2002: 8) distinguishes between 'boundary' and 'border'. The boundary, he suggests, is 'a strict line of separation between two (at least theoretically) distinct territories'; whereas a border is 'an area of interaction and gradual division between two separate political entities'.

Political geographers have traditionally understood borders as key to establishing and defending the territory of the nation-state. The modern nation-state arose in Western Europe, and one of its most important defining factors was a clearly delineated border. In order to define membership or citizenship and attach this belonging to a defensible territory, states rendered national landscapes legible with the use of borders. Passports were developed in order to define and regulate the mobility of those populations residing inside and outside the nation-state (Torpey 2000). Clearly delineated borders thus came to replace the more ambiguous and fluid boundary zones that once existed between populations. Because these international borders are a relatively recent invention, they are often contested and subverted by nations older than nation-states. The borders of some African states, for example, confirm the practice of Western European colonial powers drawing lines far away from their operationalization on the ground, where they actually cut across tribal nations. In spite of this mismatch and the territorial conflicts that often result, these borders remain straight.

Thus the border is a clear location where nation-states attempt to differentiate places and peoples; states 'see' (cf. Scott, J. 1998), police and regulate borders in all sorts of ways. States exercise the right to control the movement of goods and persons across borders.

Political geographers have studied geopolitical relationships between states. They inquire into the role of border disputes as expressions of international relations between states (Agnew 1998). They also understand the boundaries as expressions of lines drawn by those who engage in geopolitical relations: 'for hundreds of years geopoliticians have drawn lines of inclusion and exclusion that were based on power politics, culture, and even physical geographical arguments' (Paasi 2001: 9).

Meanwhile, thousands of individuals embark on distinct kinds of border crossings around the globe every day. The border is a site experienced in a variety of ways, sometimes painfully and violently, particularly for those fleeing political persecution in search of protection away from home, and other times with relative ease, such as by those who cross daily to work.

Some political geographers have written recently of the hyper-policed, militarized border in the geographical imagination, as in the case of the Israel–Palestine border and the Berlin Wall. These borders serve as dramatic and sometimes deadly sites where sovereign power is exercised. In the case of Palestine, borders are hyper-regulated through a series of onerous checkpoints where identity documents are colour-coded in order to control mobility along short distances, from one town to the next.

Still other borders can be mapped in a more fluid fashion, as borders in motion. Knowing that their locations are often disputed, it is important to ask where borders are located and how their governance is changing. International travelers often cross borders in airports. With globalization, the border has moved into our daily lives. Some workers cross borders daily to work, whether in the high end service economy of the financial world or the low end service economy as construction crews and domestic workers. People can now have dual citizenship and even carry multiple passports (Ong 1999).

These distinct kinds of border crossing, ranging from the mundane to the deadly, expose power relations, rendering visible power that often operates in a more subtle fashion. People who have more money in the bank, who drive nicer cars or can afford airplane tickets and tourist visas, will have a better chance of crossing successfully than those who are poor and may attempt to cross on foot or by bus or boat, or in an illicit fashion, undocumented by the state.

Borders have taken on full conceptual lives for political and cultural geographers in recent years. They do the work of delineating not only space, but identities in space. The mere existence of a border creates two dualities: us and them, inside and out, domestic and foreign, citizen and alien, legal and illegal.

National borders are sites where individuals must express their identities in the legal terms of citizenship. When crossing, they are literally rendered visible, legible to the nation-state. Anthropologist Michael Kearney (1991: 58) notes, 'It is in this border area that identities are assigned and taken, withheld and rejected. The state seeks a monopoly on the power to assign identities to those who enter this space'. Thus the border is not only a site of division, containment and crossing, but an exercise linked to processes of identity formation and identification.

Who are we, and how does that relate to our location of birth, work and residence? Where do these locations begin and end, and where do they blur? The border functions as both a material and metaphorical separation. Borders exist not only to divide but as sites of transgression, subversion and resistance. Struggles over identity and belonging arise there.

Many examples throughout history offer stories of such transgression, from the racially integrated freedom bus riders who rode through and challenged racially segregated parts of the southern United States during the civil rights movement of the 1960s to anti-detention activists who pushed down walls around detention centres where asylum seekers were detained in rural Australia at the turn of the twenty-first century (Mares 2002; Mountz, forthcoming).

Borders represent not only sites of policing and transgression, but of deep ambiguity, as in the case of migrant workers. Migrant workers often have temporary legal status in the nation-state where they have migrated to work; they have the right to work but not to remain. As such, global inequalities are inscribed, written on to the bodies of foreign workers whose identities are reproduced at work, at home and at the border (e.g. Pratt 2004; Wright 2006). Because of the ways that migrant workers from the global South are often racialized in their daily lives in the global North, the border migrates in their lives, moving with them and perpetually reconstituted in their interactions with employers and others. Geraldine Pratt (2004), for instance, demonstrates the ways that national immigration policies are scripted onto the bodies of Filipina domestic workers through Canada's Livein Caregiver programmes. (For more information on immigration policies see Chapter 15 on migration.) Let's turn to two more specific examples now, that illustrate how borders work not only as fixed lines that demarcate the territories of nation-states, but as powerful forces in the daily lives of the people who live in those territories.

Case Studies: A Rigid Border and a Border in Motion

Case Study 1: Gloria Anzaldúa and the US–Mexico Border

The US–Mexico border occupies a prominent place in the geographical imagination and public discourse of Mexican and American citizens. Gloria Anzaldúa is a Chicana writer and poet who grew up in the border region in south Texas. Her text, *Borderlands/La Frontera* is an autobiography and conceptual tome on life and loss in the borderlands. She presents the border as an edge, the end of one thing and the beginning of another, 'a dividing line, a narrow strip along a steep edge' (1987: 3). This border divides people experiencing great socio-economic disparities: 'The US–Mexico border *es una herida abierta* [an open wound] where the Third World grates against the first and bleeds' (1987: 3). Anzaldúa renders the border a living, breathing entity through her use of metaphor, poetry and mythology. She grew up along the border in Texas's Rio Grande Valley where the US–Mexico border acted as a permanent presence in her daily life. She adeptly demonstrates the ways in which we internalize borders that then shape our daily identities, regardless of where we live. She also argues that individuals fight the power of the border to impose, regulate, and colonize. As such, border regions offer opportunities alongside barriers. The person who straddles cultures and contradictions along borders will have the ability to challenge binaries with what Anzaldúa calls a 'consciousness of the borderlands' (1987: 77): the ability to operate in more than one culture, more than one language, to inhabit more than one identity.

Anzaldúa shows that borders divide people not only materially, but also metaphorically by reinforcing difference: 'Borders are set up to define the places that are safe and unsafe, to distinguish us from them'. While she characterizes the border as itself an 'open wound', Anzaldúa suggests that borders simultaneously scar the bodies of those who inhabit border regions:

> 1,950 mile-long open wound
> Dividing *a pueblo*, a culture
> Running down the length of my body
> Staking fence rods in my flesh
> Splits me splits *me raja me raja*
> This is my home
> This thin edge
> Of barbed wire (1987: 2–3)

She draws a parallel between the marginalization of inhabitants of border regions that exist along the edges of sovereign territory and those who occupy zones marginalized from 'mainstream' society because of their distinct locations, histories and identities: 'the perverse, the queer, the troublesome ... in short, those who cross over, pass over, or go through the confines of the "normal"' (1987: 3). As a queer woman whose sexual preferences are not normative, as a woman of colour, a bilingual speaker of Spanish and English, Anzaldúa continuously writes her own location in the borderlands long after she leaves the actual US–Mexico border region.

Case Study 2: The Border in Motion

Borders themselves are evolving in geographically imaginative ways. The policing of borders is growing more integrated through something called *harmonization* and *perimeter theory*. Members of the European Union, for example, harmonized regulatory controls over migration, opting to dramatically loosen control over internal migration among member states but to strengthen regulation of the perimeter of the newly integrated European Union. This integrated border-policing manifests in the joint policing agency called Frontex, formed in 2004. In North America, Canada and the US collaborate in a more subtle fashion, such as sharing-information through integrated databases to build a perimeter by regulating movement to North America.

Collaborative policing efforts serve as reminders that borders are not only fixed in time and space, but also dynamic entities, always in motion through the very people who police them. Alison Mountz (2006) has studied the deployment of civil servants abroad by Canada, the United States, Australia and the United Kingdom. These civil servants are called airline liaison officers. They work in foreign airports where they do not hold jurisdiction but informally police travelers in queues to board international flights. They are looking for false documents or 'false' pretences for travel, such as a person who is traveling on a tourist visa but plans to make an asylum claim on landing on sovereign territory. The stopping of these persons before they arrive is a practice called 'interdiction'. These officers also train airline staff to recognize false documents. They share information on human smuggling rings with local authorities, and call ahead to authorities on the ground in the 'host' country when an individual they suspect might be traveling under false pretences boards a plane. Through these programmes, called 'the multiple borders strategy' by Canada and 'Global Outreach' by the

United States, civil servants are extending the border outward well beyond traditional sovereign territory to police what civil servants call 'hot spots' where human smuggling operates (see Coleman 2007; Mountz 2006). In these cases, the border itself is in motion through the daily efforts of authorities to police and patrol migration. It follows migrants well beyond traditional spaces of sovereign territory.

Such policing strategies are controversial because they preclude people's chances of reaching sovereign territory to make an asylum claim. Another example of the ways that civil servants are pushing the borders of sovereign power offshore through the controversial extension of border enforcement is Autralia's 'Pacific Solution', in place since 2001 (Mares 2002; Taylor, S. 2005). In 2001, Prime Minister John Howard built a national re-election campaign on anti-immigrant sentiment that targeted people who were trying to reach Australian sovereign territory by sea with the assistance of human smugglers. Howard led an aggressive enforcement regime, which involved detaining migrants that Australian authorities intercepted at sea in detention centres on islands, including Melville, Nauru and Christmas Island (see Figure 17.4). Those who were detained on islands were not allowed to make asylum claims. Some of these islands, such as Nauru, were not Australian. Others were Australian (e.g. Melville and Christmas Islands), but the government excised them from the 'Australian migration zone' – a territorial categorization from which asylum applications *could* legally originate. By using geography creatively, the Australian federal government was able to distance potential asylum seekers from sovereign territory, thereby making an asylum claim impossible. Part of the 'Pacific Solution' was the 'power of excision'. The Australian Parliament met in 2001 to retroactively declare those sites where migrants had landed no longer part of Australia for the purposes of migration. The ability of a nation-state to remove parts of its own territory for some individuals, but not others, serves as a dramatic expression of the mobility and power of borders and their intimate ties to legal status, sovereign power and identity (Mountz 2004).

Future Research: Disorder and Borders in Motion?

As borders become less fixed in the global era, political geographers are growing less interested in the ordering of the world, and more interested in understanding disorder in the world (Agnew 2002; Flint 2002; Gomez-Peña 1996). The same can be said of scholarly interest in and

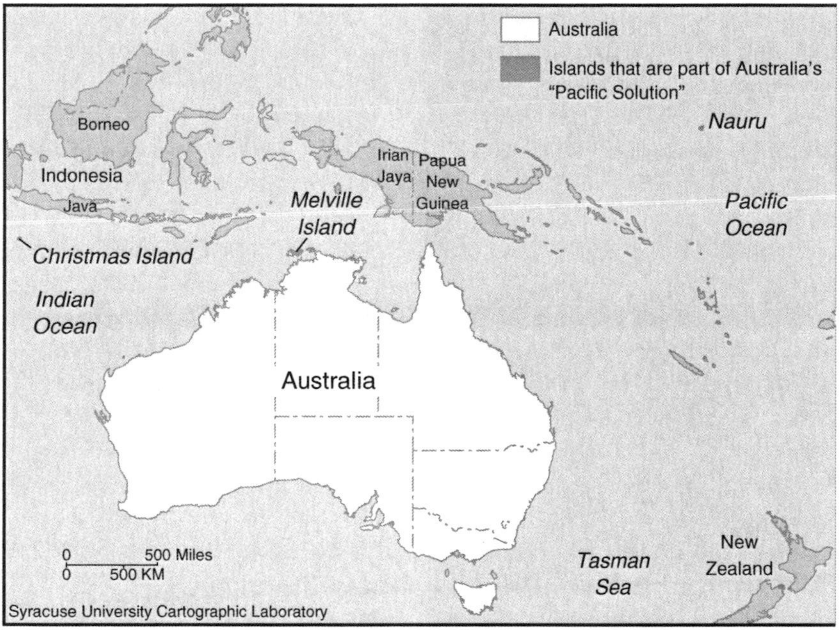

Figure 17.4 Australia's 'Pacific Solution.' Christmas Island, Nauru, and Melville Island were used to detain asylum seekers caught at sea. Asylum seekers could not make asylum claims in Australia from these locations. Nauru is not part of Australian territory, and Melville and Christmas Islands, which are part of Australian territory, were removed from the Australian migration zone

conceptual understandings of borders. How will old ways of studying conflict between states be redrawn alongside the reconfiguration of the border? If borders are in motion, what will be the new 'theatres of war' where conflicts between states unfold? Will new forms of warfare mimic the practices of terrorist groups where combatants straddle international borders? How will borders be utilized to police not only conflict but membership in the form of citizenship?

In a more chaotic and immediate, intimate and global world, borders are no longer clearly delineated lines. Louise Amoore (2006), for example, writes about the biometric border. As the name suggests, the biometric border regulates crossings through the collection and measurement of biographical data such as retinal scans and digitized fingerprints at ports-of-entry where citizens and non-citizens enter nation-states. These data are catalogued and shared among government agencies through integrated databases. The database allows a border guard to access a person's history of border crossings with the click of a button. Unlike the borders that cartographers have traditionally drawn on maps, the biometric border is reconstituted in a more dispersed and transnational fashion enabled by advances in technology.

To police increasingly fluid borders, the state, its civil servants and its technologies must adapt to borders in motion as well as an increasingly sophisticated human smuggling industry that also continuously adapts to advances in enforcement. In short, an infrastructure is evolving to support the policing of the border. This involves integrated technologies such as databases and software that enable authorities at the border to know in a moment where and when a person has crossed other international borders. Retinal scans, heat sensors and integrated databases take the place of guard booths.

These changes in the regulation of borders connect with broader changes in practices of governance. Michel Foucault (1977, 1991) identified these changes as a historical shift from sovereign to disciplinary power; a move from visible and violent impositions of power associated with a sovereign leader to daily practices of power associated with the circulation of intimate knowledge. As nation-states grew and bureaucratized, sovereign power was exercised in the form of census-taking, counting the population, gathering information and regulating citizenship. As states amass more biographical data about populations over time, individuals and communities begin to self-police in response to practices like racial profiling and required registries, avoiding public sites where surveillance takes place. There is, therefore, an in increasingly important and intimate relationship between border and body (Mountz, forthcoming). What theories will future political geographers devise to understand these borders?

KEY POINTS

- A border distinguishes between two separate entities. It is often materialized on the ground in the form of a fence, hedgerow, sign or checkpoint.
- For political geographers, the border is often international; it separates the national territories of two nation-states.

- Establishing and maintaining borders are crucial to the integrity of the modern state. States use a variety of mechanisms to enforce their borders, including checking travel documents, limiting border crossings to select travelers, and in some cases building obstacles to prohibit unauthorized crossings.
- National borders are sites where individuals must express their identities in the legal terms of citizenship. As such, the border is not only a site of division, but also an exercise linked to processes of identity formation and identification.
- Struggles over identity and belonging often occur in border areas.

FURTHER READING

Anzaldúa, G. (1987) *Borderlands/la frontera: The New Mestiza*. San Francisco, CA: Aunt Lute.

Mares, P. (2002) Borderline. Sydney: UNSW Press.

Nevins, J. (2002) *Operation Gatekeeper: The Rise of 'Illegal Alien' and the Remarking of the US – Mexico Bounary*. New York: Routledge.

Sibley, D. (1995) *Geographies of Exclusion*. London: Routledge.

Foucault

1. Hierachic possible obseration
2. Normalization of behavior
3. Examination

Governmentali

18 REGIONALISM

Carl T. Dahlman

Definition: Politics Above and Below the Nation

In political geography the term 'region' refers to two basic types of area defined by their historical, cultural, economic, social or political distinction from surrounding areas, but which do not typically coincide with formal political units such as the nation-state. The first are world regions, such as the Caribbean basin or South Asia, organized loosely around rather arbitrary geographic subdivisions, such as continents, but which were given supposed coherence within the racial geopolitical vision of European imperialism. More recent geopolitical projects have recast world regions as regional collections of states divided according to Cold War divisions (First, Second and Third Worlds), relative socio-economic development (global North and South), or newly formed regional associations, such as the European Union (Figure 18.1).

The second type of region is typically smaller in size and is commonly, if not always accurately, thought of as sub-national. These regions are identified with considerable historical, cultural or economic internal similarity and they are frequently associated with strong local, minority or ethnic identities. Such regions may comprise relatively compact even insular areas as in Aceh, Indonesia or East Timor, while others like Catalonia and Brittany claim large areas within well-stablished states. As a political concept, regionalism refers to the structure and mobilization of a regional identity built around the particularities of these smaller areas, not unlike the way that nationalism is seen through the lens of its claims to a homeland. (For more information on nationalism see Chapter 23.) At the same time, regionalisms are distinct from nations and, regardless of whether they inform integrationist, autonomous or separatist politics, they are formed in a relational binary to more dominant political identities (Figure 18.2). Regionalisms are often associated with marginalized peoples and areas within the modern state, but not always. England and France, for

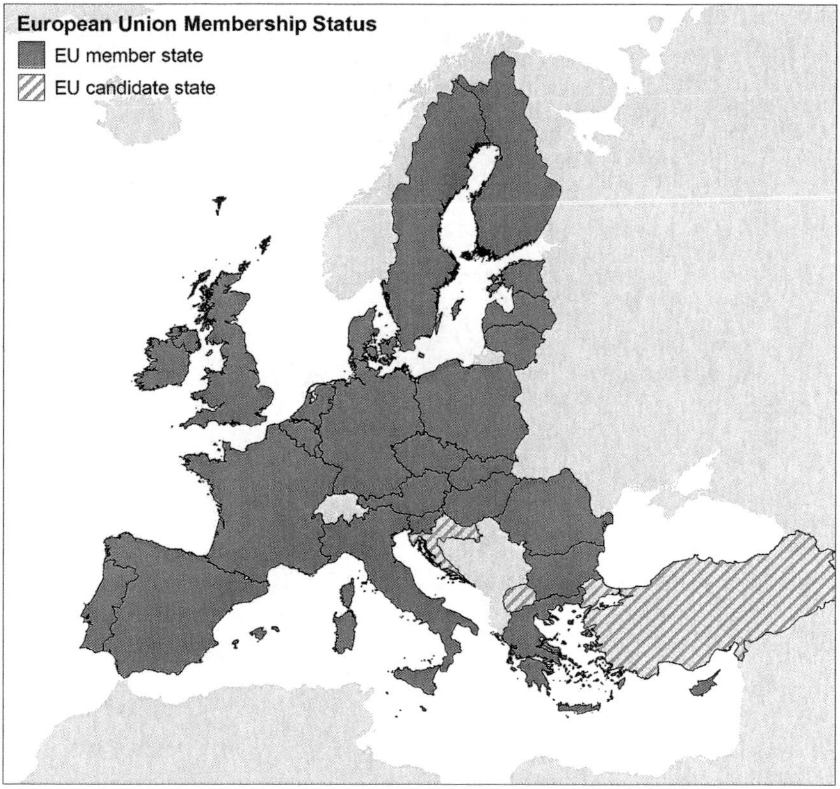

Figure 18.1 European Union Member and Candidate States (Map by and courtesy of Carl T. Dahlman)

example, are nations built around particular regional identities in, respectively, London and its home counties and the Paris basin, even if the ruling elites have long denied their regional biases (Jones and MacLeod 2004; Loughlin 1996). In fact, more marginal regions are often energized by a sense that the state is managed by and for the benefit of a few regions at the expense of others. Such differences vary in their importance across historical and geographical contexts, but they draw into sharp relief the constructed aspects of the nation-state ideal. (See Chapter 1 for details on the concept of the nation-state.)

Regionalism has often been viewed by state-building elites as traditionalistic, backward or reactionary, while regionalist movements

Figure 18.2 Provinces claimed in whole or in part as Basque country lie on both sides of the French-Spanish border (Map by and courtesy of Carl T. Dahlman)

typically claim real and perceived historical grievances. Contemporary regionalist motivations, however, are more complex and geographers attribute their emergence within the modern state system to a variety of causes (Agnew 2001; Keating 1997). First, the process of decolonization and the collapse of the Cold War have brought regionalist complaints to the fore. These changes in the geopolitical order, of course, had effects within both new and old states where the legacies of colonial borders, under-investment and expropriation, or the newfound success of particularistic identities, can trigger long-standing or new concerns in certain regions. The Acholi region of northern Uganda, for example, was an area of labour and resource extraction that benefited the south of the country under British rule. The long-suffering north eventually gave rise to an Acholi militia that briefly took over the capital in 1985, only to be driven back a year later, initiating two decades of armed rebellion within northern Uganda (Mutibwa 1992).

A second cause relates to the challenges of economic competition in an age of globalization. This is especially problematic in states going through neo-liberal transformations meant to enhance the economic advantage of more competitive regions while cutting government spending on less developed ones. The success of central and northern Italian industrial districts has been enhanced by protectionist trade schemes, for example, while growing pressure on government spending threatens to leave the less well developed south of the country further behind. A third, somewhat parallel cause is the growth of 'cultural' responses to colonialism and globalization, seen in an assertion of local identity around regional customs and collective memories, even as those traditions are changed or invented in the process. This may be seen in the 'invented traditions' of Scotland, most popularly the kilt and tartan, which have become symbols of Highland regional, even Scottish national, difference from England (Trevor-Roper 1983).

A fourth account of the emergence of regionalist movements relates uniquely to political developments in Europe. Modern state-building in nineteenth-century Europe sought to diminish the importance of regions and regionalisms that challenged the primacy of a centralized state. Thus, for example, did France seek to diminish regional differences by cultural homogenization, primarily via French language instruction. Largely bypassed by modernizing forces, these marginal regions after World War II became targets for economic development and incorporation into the state. In response to the challenges of globalization, regionalist movements emerged in the 1980s as the locus for greater regional control over cultural, political and economic policy (Keating 1997). Europe today is a patchwork of regions with very uneven significance. Germany, for example, is comprised of 13 states, two city-states, and one state of both a city and an 'area'; nearly all were separate political units prior to German unification and they retain varying degrees of regional identity. In contrast, Belgium is made up of two fairly distinct linguistically defined regions: Flanders and Wallonia. The United Kingdom has marked tensions between historical regional identities, such as in Cornwall, and bureaucratic regional planning regions that are a creation of the central government (Jones and MacLeod 2004).

European integration has added fuel to some regionalist movements that seek greater representation and voice in EU decision-making. Since about 70 per cent of EU legislation is implemented at the regional or local level, their concerns have been taken somewhat seriously. An effort to bring subsidiarity (an EU principle meant to bring decision-making as close to the citizens as possible) to the regional level produced the Committee of the Regions, a consultative body to the EU created by the 1994 Maastricht Treaty. While the Committee provides a venue for regional voices, critics often describe it as toothless because it has no independent policy-making authority. The state-centric norm that still exists in the EU, moreover, means that real regional participation in EU policy-making varies according to the role of the regions in each member state. Some regionalist authors and regionalist movements envision a federalist Europe in which the member states are effectively made redundant, surpassed by an institutionalized 'Europe of the Regions' (Loughlin 1996). Still other regionalist movements reject both the state and the EU, seeing neither as a legitimate political entity. Bosnian Serb nationalism, for example, largely rejects their state, Bosnia and Herzegovina, as well as European Union involvement there and in the surrounding region.

Regionalisms operate in dramatically different contexts, directly affecting their political outcomes. In Europe, regional autonomy or minority rights movements often ground their legitimacy in historical territorial claims while using the existing state administrative apparatus to push for accommodation. More independence-minded groups, including parts of the Republican (Northern Ireland) and Basque (Spain) movements, have used violence where accommodation is either not offered or not sufficient. Regionalisms in parts of the post-colonial world, by contrast, usually do not have recourse to accommodation through regional or local administrative procedures. These weakly centralized governments often view any regionally based movement as a challenge to state authority, since there exist few other common social institutions that tie these polities together as, for example, in Darfur and the southern regions of Sudan. The outcomes for regional movements – whether they are accommodated or repressed – are driven by a political and economic context although most states prefer the status quo of centralized power and regional marginality. State anxiety over secessionist or disloyal regions often brings military repression, which illicitly funded regional actors may be able to fight off (Le Billon 2001). This has been the case in northern Iraq where Kurdish political parties were able to raise cash during the 1990s by exporting oil. This dynamic has amplified since the US invasion of 2003, making it more likely that the Kurds will seek absolute regional autonomy or even independence should the central government fail.

Evolution and Debate: Questions of Process and Location

Current thinking about regionalism is extensive and diverse, but two overarching conversations are present in much of the contemporary literature. The first centres on questions of process. The second is focused on where regions are located and are likely to appear in the future.

Questions of Process

Process-related questions centre on identifying and analysing factors that shape and sustain regionalisms. Two different intellectual traditions inform our understanding of this question: those that approach the region as a historical cultural area and those that examine it as primarily an economic functional area. The first draws from an older tradition of studying regions by dwelling on the historical, cultural, political and economic features that give them their distinctive character. As a form of contemporary scholarship, this approach took shape in Vidal de la Blanche's studies of how traditional livelihoods and regions in France were being transformed by modernizing influences at the turn of the last century. His approach is echoed in the Anglo-American tradition of descriptive, humanistic writings about regions (Gregory 2000). In practice, regionalist movements also make use of arguments about the distinctiveness of their respective historical cultural areas but through interpretations that invest regional identities with contemporary political values while bolstering their assertions of political legitimacy. Noted Finnish geographer Anssi Paasi cautions that academic and institutional forms of talking about the identity of a region are themselves political acts, that are different from but implicated in how people identify themselves through regional political, social and economic institutions (Paasi 2003: 478).

The other intellectual approaches to regions and regionalism are based on analyses of the uneven economic development commonly found between regions. Regional analysis of this sort gained popularity in the

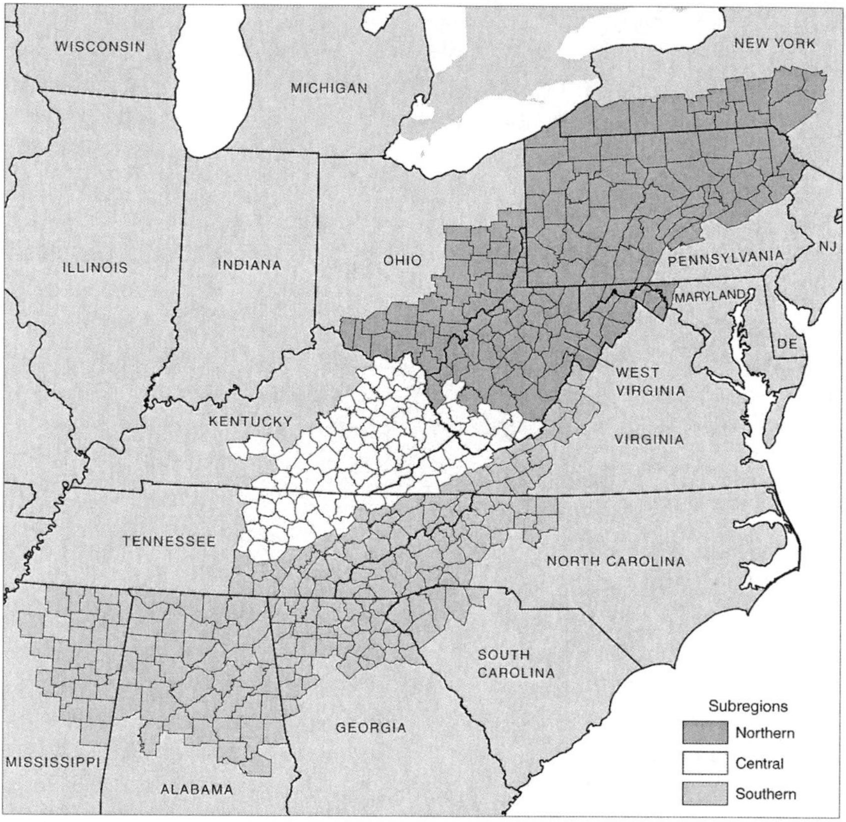

Figure 18.3 The Appalachian Regional Commission comprises counties, many with low socio-economic averages, targeted for government-sponsored development programmes (Map by and courtesy of Carl T. Dahlman)

mid-twentieth century and relied on heavily quantitative inquiries to identify the significant factors in regional economic performance. Geographers and their mathematical models were deeply invested in the production of official policy and were themselves producers of regional classifications based less on existing cultural regions than on administrative geographies of bureaucratic convenience. An interesting example is the Appalachian Regional Commission, a 1960s bureaucratic invention intended to direct federal assistance toward regions with marked social problems, but which also contributed to Appalachia's depiction as a cultural region (Figure 18.3). The geography of regional economies provided a framework for understanding how increasingly mobile global economic forces might make regions, not states, the locus of politics (Scott, A.J. 1998). As this 'new regionalism' concept was laid on top of existing regionalisms, it created novel political responses in trying to capture development capital. Regional political claims of economic marginalization gained new legitimacy in some cases, yet the resulting policy choices often favoured reinvestment in existing core regions.

Locating Regions

The second overarching conversation in contemporary scholarship about regionalism is the question of where we can locate the region and how it relates to other scales. Obviously, the different ideas of what defines and sustains regionalisms provide rather different geographies of regions themselves. Cultural-historical approaches to regionalism tend to focus on ethnic or would-be national territorialities that are supposed to pre-date existing political entities. These regions are often written about as though they were wholly nested within a state, but culture-based regionalisms often stretch across state borders, as with the Slovene regions in Italy and Austria (Herb and Kaplan 1999). And although these 'ethnic' regionalisms exist in all parts of the world, they are more likely to be the object of concern in developing or conflict-prone areas than in highly developed states. In contrast, the urban space of the 'new regionalism' is perhaps too often associated with sites of high-modernism like central business districts or technology parks. To be sure, competitive urban regions are the pre-eminent site for the networks of finance, innovation, research, development and technology necessary to sustain flexible production and knowledge-based economies (Ward and Jonas 2004). New regionalism is thus associated with spatial shifts in the leading edge of capitalist accumulation strategies, emphasizing new technologies and consumer markets that characterize our understanding of globalization in narrow terms. We should not rule out, however, urban-centred regionalisms in developing spaces, no matter how tenuous their role in a globalizing, but not yet global, economy (Mbembe 2000).

Case Study: Kurdish Regionalisms

Examining the political geography of the Kurdish population in the Middle East through the lens of regionalism would seem to deny a long-standing truism that 'the Kurds are the largest nation without a state'.[1] The area in which Kurdish populations have been significant, if not predominant, extends from the mountainous regions of eastern Turkey and northwestern Iran into northern Syria and northern Iraq, extending southwards along the Zagros mountains. Kurdish nationalists in the twentieth century have often claimed larger areas based on historical and demographic suppositions, but nowhere have the Kurds established a meaningful polity that embraced this larger realm.[2] The fact remains, however, that Kurdistan was divided by the boundary-drawing after World War I. Kurdish regions within each state, therefore, have experienced often tense relations with the central governments in Turkey, Iraq, Iran, Syria and the Soviet Union. Kurds are culturally different from the regional identities that predominate among the ruling elites in each of these states, and Kurdish areas were largely ignored in social and economic development programmes, if such initiatives existed at all. For the last hundred years, Kurdish regions have remained poor and underdeveloped in comparison to other parts of these countries.

Taking a closer look at each of the Kurdish regions, we can see how a relatively homogeneous if rather weak nationalist movement has over time become four different regionalist movements. A putative fifth Kurdish regionalism in the southern Caucasus has no identifiable form, although self-identified Kurds continue to live there. The largest Kurdish region is that in Turkey, and comprises about 13 million Kurds or about half the total Kurdish population. Since the formation of the modern Republic, the Kurdish region in Turkey has remained an exploited, marginal area, passed over by the modernizing industrialization of western and central Turkey. The Turkish state, seeking to establish a civil identity, nonetheless constructed a largely Turkic cultural identity for the state, and ethnic identities like the Kurdish were actively repressed as both backward and a potential source of territorial disintegration or unrest.

The general misery of the Kurds in Turkey was championed by numerous Kurdish rights and Kurdish national movements at various points in the twentieth century but all were put down by the state. Most notably, in the 1980s Abdullah Öcalan led a violent guerilla movement known as the PKK (Kurdistan Workers' Party) that sought national liberation for the Kurds in Turkey. Despite the serious challenge posed by the PKK to the government, Turkey retained control over the region and eventually captured Öcalan, effectively ending the war. Turkey has also begun a series of social development initiatives in the southeast Kurdish region. However, some of these initiatives are controversial mega-dam projects that will displace many people. The state has also removed most of its restrictions on the Kurdish language and related cultural expressions, and Kurds are no longer euphemistically called 'mountain Turks'. Much of this change has come with the liberalizing influence of EU institutions as Turkey seeks membership, but what remains clear is that the Kurdish identity in Turkey is institutionally and politically quite different from Kurdish movements in neighbouring countries.

Compared to the Kurdish movements in Turkey, those in Syria and Iran are weak and poorly formed. Both Kurdish areas are at best cultural regions with little meaningful political or economic development. In Syria, massive displacement and urbanization have disrupted any *in situ* territorial claims, while in Iran the Kurds have been in a standoff with the state since the 1979 Islamic Revolution, leading to a partially successful push for greater cultural autonomy but little serious consideration of a more formal Kurdish regionalism.

The Kurds in Iraq are in a very different position than those in any other country (Figure 18.4). First, they are a compact and relatively distinct cultural region in the north of the country, comprising more than four million people. Second, Iraqi Kurds are dominated by two of the strongest parties in any Kurdish region, the KDP (Kurdistan Democratic Party) formed in 1945 and the PUK (Patriotic Union of Kurdistan) formed in 1975. The KDP maintained control over most of northern Iraq for decades before the Iraqi state declared it a Kurdish Autonomous Region in 1974. Kurdish claims to the oil-rich town of Kirkuk, among other things, led to war between the KDP and the central government, from which emerged the PUK. The KDP and the PUK tried to take advantage of the Iran–Iraq war in the 1980s to consolidate their control over northern Iraq. In its response to the Kurdish threat, the Ba'thists under Saddam Hussein launched a genocidal campaign (*al-Anfal*) against the Kurdish civilian population of northern Iraq during the late 1980s. After the 1991 Gulf War, the US and UK protected northern Iraq, in which the Kurdish parties eventually formed the Kurdistan Regional Government that has continued to operate like a highly autonomous federal region within Iraq, largely free of central control even under the new Iraq Constitution created following the US invasion.

Importantly, the Kurdish region in northern Iraq is undergoing what might be described as an economic boom with features reminiscent of

Figure 18.4 Divisions among the Kurds in Northern Iraq speak not only to the regionalization of the Kurdish issue among the states that divide Kurdistan, but also the political territorialism of Kurdish political rulers. (Courtesy of the US Central Intelligence Agency, January 2003)

the competitive city-region described by new regionalism. The major Kurdish cities are sites of rapidly diversifying economies, flush with oil revenues, which provide a zone of relative stability. If new regionalism has a vision of regions developing political-economic institutions in relative independence from the state, then northern Iraq serves as a prime example, but our high-modernist visions of competitive urban regions are perhaps too limiting to recognize this. The example of northern Iraq also suggests that economic regionalism is not independent from cultural-political regionalism but comes at the expense of a larger

Kurdish nationalism. More importantly, the influence of external actors and central state policies in each part of the Kurdish homeland has contributed to divergent regionalisms among the Kurds. In Turkey, the Kurds are increasingly supporting mainstream politics, while the state has allowed a greater space for cultural expression. This and massive emigration from the region has drawn them away from traditional bases of authority. In Iran, a more traditional ethno-political movement remains on the defensive, while in Syria the Kurdish region is vanishing under state repression and emigration. Only in northern Iraq might a Kurdish regionalism flourish, soaked in oil revenue, yet still captured by the traditionalistic ethno-political parties.

KEY POINTS

- Political geographers describe regions as sub-national or transnational areas differentiated from their larger host state(s) by regional claims to distinctive historical, cultural, political and/or economic experiences. Regionalism is a set of claims or an identity made on the basis of the region.
- Political regionalisms based on marginalized historical cultural groups have long challenged the domestic authority of centralized states.
- Economic regionalisms tend to reflect strategies of urban capital accumulation in which desire for global competitiveness challenges the regulatory role of the state.

NOTES

1 I am, in this section, following my own work, especially Dahlman 2002, 2004.

2 There was a Kurdish Republic of Mahabad in a small area of Iran following World War II, which only lasted a year.

FURTHER READING

Agnew, J. A. (2001) 'Regions in revolt', *Progress in Human Geography*, 25(1): 103–110.

Keating, M. (1997) 'The invention of regions: political restructuring and territorial government in Western Europe', *Environment and Planning C: Government and Policy*, 15(4): 383–398.

Paasi, A. (2003) 'Region and place: regional identity in question', *Progress in Human Geography*, 27(4): 475–485.
Ward, K. & Jonas, A. E. G. (2004) 'Competitive city-regionalism as a politics of space: a critical reinterpretation of the new regionalism', *Environment and Planning A*, 36(12): 2119–2139.

Part V
Violence

INTRODUCTION

Carolyn Gallaher

By most measures the twentieth century was the most violent century on record. The formalization of the state system during the twentieth century was predicated on the monopolization of the means of violence at the state level. Indeed, the twentieth century saw the formation of dozens of standing armies. Over time, states developed ever more sophisticated weapons for their use. And capitalists played their part, industrializing the production of arms by mass-producing AK47s, grenade launchers and even aircraft designed to launch bombs and other projectiles. While it is too early to tell if the twenty-first century will be equally violent, many of the characteristics that made the twentieth century so violent continue to exist. The nature of violence may change, however, as the state system is eroded in many parts of the globe.

In this part of the book four concepts related to violence are covered. In the first chapter the concept of **conflict** is covered. In geography, conflict has historically been examined *vis-à-vis* wars between states. Increasingly, however, it occurs within states or in largely stateless areas across the globe, and involves non-state actors, such as guerilla or paramilitary groups. Conflict may also involve the study of largely non-violent contestation. Some geographers examine how different identity-based groups deal with conflicting interests. In Chapter 20 the concept of **post-conflict** is detailed. When conflicts end, violence usually ebbs but does not stop altogether. Geographers examine how violence changes after a conflict ends, with a particular focus on any changes in the spatial patterns of violence.

In the next chapter a tactic of conflict – **terrorism** – is discussed. Terrorism is a loaded term in that there is no readily agreed upon definition of the word. At a base level, geographers (and other social scientists) view terrorism as a tactic of war designed to intimidate a civilian population. While terrorism is often described as a random event, geographers note that there are distinct spatial patterns in the distribution of terrorist attacks.

In the final chapter of this part a particular type of conflict – **anti-statism** – is considered. Anti-statism refers to internal assaults (verbal or physical) by citizens on the state in which they live. The goal of anti-statism is to dismantle the state. However, unlike a traditional guerilla war where the goal is to replace one state structure with another, anti-statists usually want to divest a state of power and move it to smaller scales, such as provincial or even local jurisdictions.

Conflict can and does occur at various scales. The examples used in these chapters follow a similar pattern. The conflicts covered include contestations over who can march in the New York St Patrick's Day parade, battles to get the federal government out of Kentucky, and wars in the Great Lakes region of Africa. The aftermath of conflict in Bosnia is also considered.

19 CONFLICT

Mary Gilmartin

Definition: In Wartime and in Peacetime

In broad terms, conflict results from the pursuit of incompatible goals by different people, groups or institutions. Often, people consider conflict as violent, but this is not necessarily the case. Conflict can take place through peaceful means as well as through the use of force, though violent conflict often receives most attention.

When we consider conflict that takes place using force, there are a number of distinctions to be made. An *armed conflict* involves the use of force by all parties in a conflict, though this varies from an attack by a single soldier to an all-out war with multiple casualties. Conflict that involves the use of force by a state or by an organized group against civilians (such as massacres and genocide) is sometimes described as one-sided violence (Eck et al. 2004). *Terrorism* is also a form of conflict. Terrorists hope to secure goals by intimidation, by using or threatening to use violence against governments, groups and even individuals. (For more information on terrorism see Chapter 21.) Johan Galtung describes these as forms of direct violence (1990). He also identifies other forms of violence, notably structural and cultural violence. Galtung explains the distinction as follows. Direct violence involves murder, maiming, detention or expulsion. Structural violence involves exploitation or marginalization, such as through an acceptance of poverty. Cultural violence is whatever blinds us to direct or structural violence, or seeks to justify it. As Galtung explains, 'cultural violence makes direct and structural violence look, even feel, right – or at least not wrong' (1990: 291).

Within traditional political geography, particularly geopolitics, the study of conflict has focused on *wars*. Braden and Shelley (2000) identify three types of traditional war: inter-state wars (between two states); civil wars (within a state); and world wars (between blocs of states). At the beginning of the twenty-first century, there were 19 wars ongoing, the majority of which were civil wars (O'Loughlin and Raleigh

2007). However, political geographers also recognize a growing global dimension to warfare. Klaus Dodds describes it as the globalization of terror: this refers not only to multinational terror networks such as Al Qaeda, but also to the globalization of transport and communications, which facilitates the activities of terror groups (Dodds 2005: 198–200).

The focus on war in political geographies of conflict is important, but it often comes at the exclusion of other forms of conflict, such as interpersonal conflict, which may or may not be violent. The focus on war also tends to preclude analysis of how to prevent or overcome conflict. It serves to provide a negative understanding of conflict, even though conflict may also ultimately be a catalyst for beneficial change.

Political geographers have traditionally focused on three main causes of conflict – territory, ideology and resources. Many conflicts are caused by a combination of these factors. More recently geographers have identified ethnicity as a common cause of conflict, particularly in civil wars, although some, like O'Loughlin (2005b) suggest that economic factors generally underpin most ethnic conflicts. In recent years, political geographers have also paid more attention to identity-based and place-based conflicts, which sometimes occur at a smaller scale, and with less violence, than the wars that have traditionally been the focus of interest and research. As examples, Brown et al. (2004) discuss 'culture wars' in the context of failed attempts to appeal against a gay rights ordinance in Tacoma, Washington; Shirlow (2006) discusses ongoing contestation in 'post-conflict Belfast'; and Marston (2002) discusses conflicts over the New York St Patrick's Day Parade discussed later in this chapter). However, with this changing focus comes an apparent reluctance to use the term 'conflict'. Instead, the term 'contestation' is often used to describe more locally-based and identity-based disputes, while the term 'conflict' is most often used to describe state-centred disputes.

Evolution and Debate: Conflict or Contestation?

Early studies of conflict within political geography often centred on war, for example by providing guidance on strategic planning for the purposes of war. Work by Halford Mackinder in the early twentieth century introduced the concept of the Heartland, the part of the world that would need to be controlled in order to ensure global control and dominance. This focus has continued, with geographers providing

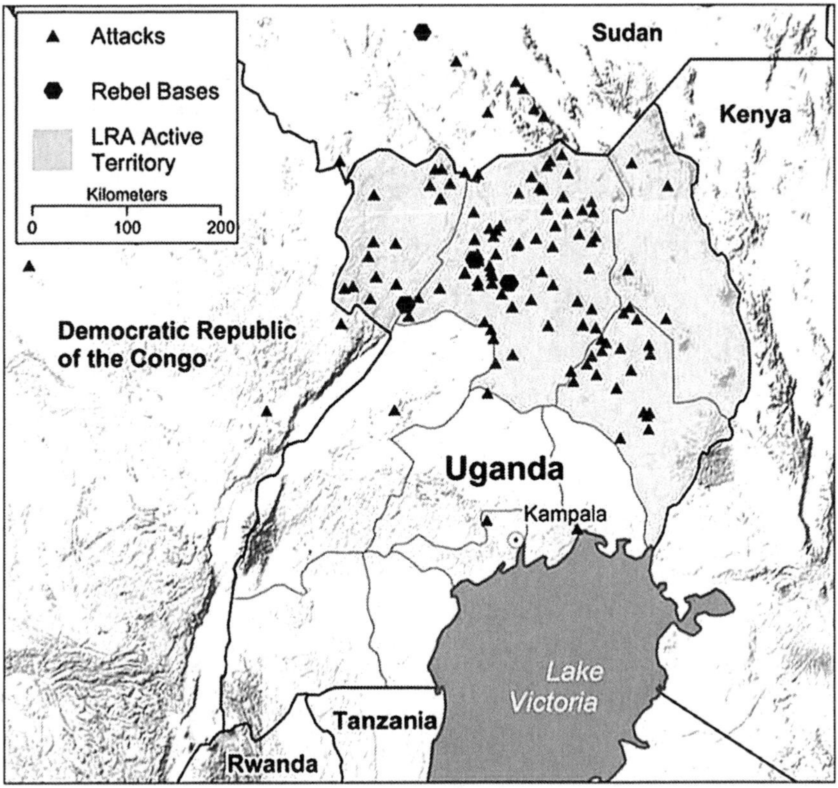

Figure 19.1 Active Territories for Lord's Resistance Army (1994–2004). (Reproduced with permission from Cox et al. 2007)

insights into tactical as well as geopolitical strategies (O'Loughlin and Raleigh 2007). However, while many political geographers served the cause of war, whether as theorists or practitioners, many others were concerned with using their knowledge to show the effects of war. O'Loughlin and Raleigh highlight the ways in which geographers have increasingly provided critical views of war and the military, ranging from mapping the effects of bombing to constructing maps of global conflicts (O'Loughlin and Raleigh 2007) (and see Figures 19.1 and 19.2). Geographers have also studied the aftermath of conflict, both symbolically and materially. This includes work on the landscapes of postnationalist and post-colonial states (see Nash 1999; Whelan 2003, in relation to Ireland).

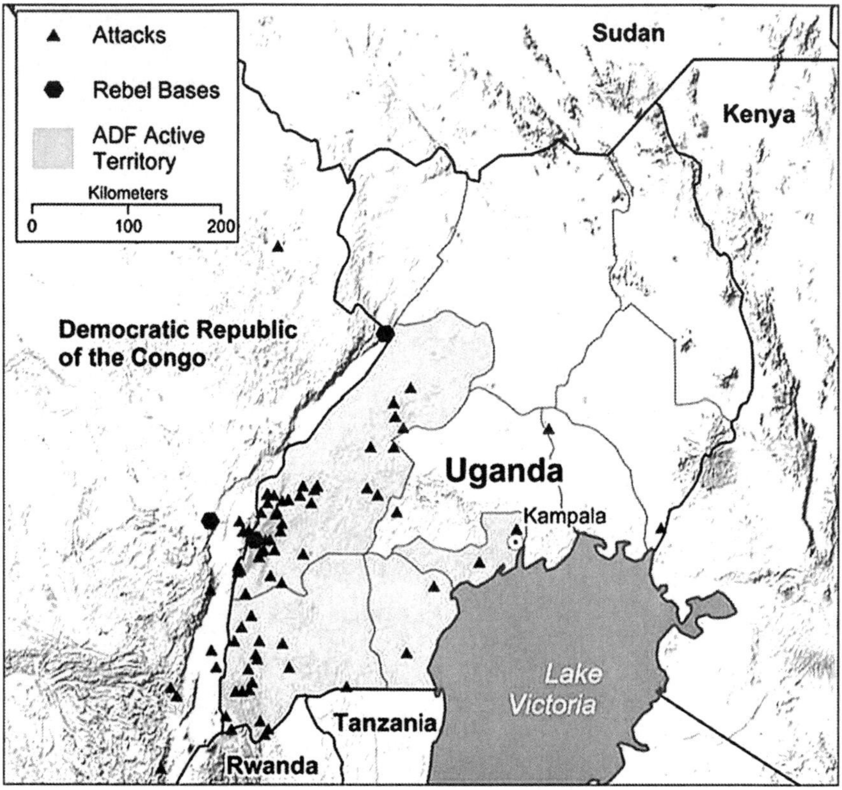

Figure 19.2 Active Territories for Allied Democratic Forces - ADF (1996–2002). (Reproduced with permission from Cox et al. 2007)

The growing interest in identity politics in political geography since the 1980s has allowed for a more nuanced analysis of conflicts over ideology, particularly ideologies related to social differences such as gender, race, ethnicity or sexuality. However, because these studies fit within the 'small p' tradition of politics, they are often defined as cultural or social geography rather than political geography. As a consequence, identity conflicts are regularly described as contestation, which serves to reinforce the conceptualization of conflict as in some way state-centred. Despite this categorization, attention to identity has significantly expanded the scales at which conflict is investigated and theorized. Now, conflicts may be seen as personal, as domestic, as street or locality or city based. They may also be seen as international or transnational, operating beyond the immediate control of the state, though often influenced by state concerns.

The theorization and study of conflict within political geography has been reshaped more recently by the attack on the US on 9/11, 2001, and by the subsequent wars in Afghanistan and Iraq. This has led to a renewed focus on war, and much recent work within political geography has attempted to make sense of the events and their resulting geographies. The so-called 'War on Terror' has given rise to numerous articles, special issues of journals and books on the topics (for example, *The Arab World Geographer*, 6(1), 2003; Ettlinger and Bosco 2004; Gregory 2004). At the same time, geographers have called for more critical attention to military geographies and the militarization of geography. Woodward highlights the strong relationship between military geography and war, commenting that 'military geography underpins military violence' (2005: 724). She argues for an approach to military geography that examines more than armed conflicts and that is critical of the ways in which military activities shape everyday lives (Woodward 2005).

In addition to these two significant shifts in the theorization of conflict, another area of interest has emerged. This is the study of resource conflicts, whether these conflicts are between or within states, involving non-state or supra-state actors, or are occurring at the local, national or international scale (Laurie 2007; Le Billon and El Khatib 2004). Often, this is framed by the broader study of political ecology (see Peet 1998: 95–97). Le Billon has suggested that approaches to the study of resource conflicts have shifted from strategic or scarce resources to the variety of conflicts that occur over resources (Le Billon 2007).

Case Studies: St Patrick's Day Parades in New York; the Narmada Valley in India

The issue of identity and conflict over identity is at the core of the first example, Sallie Marston's account of St Patrick's Day parades in New York City (Marston 2002). Since 1991, the annual St Patrick's Day parade in New York has been a site of conflict between the parade organizers, the Ancient Order of Hibernians (AOH) and the Irish Lesbian and Gay Organization (ILGO). The ILGO, formed in 1990 by recently arrived Irish immigrants to New York, applied for permission to march in the 1991 St Patrick's Day parade. The application was refused, ostensibly on the grounds that it was late. However, refusals in later years suggested that the ILGO was being excluded from participation in the parade under its own banner because its articulation of Irish identity was at odds with that of the parade organizers. The Catholic religion is central to the beliefs and practices of the AOH, many of whose members are older and more conservative. Marston suggests that for these members of the AOH, 'it is one thing to be Irish and homosexual in private; it is quite another to be public about that identity' (2002: 389). Similar conflicts took place in other US cities, such as Boston, where the parade organizers banned the Gay, Lesbian and Bisexual Irish Group of Boston from taking part, a decision upheld unanimously by the US Supreme Court. In this way, the annual public performance of Irish identity in the US through the St Patrick's Day parades is a site of conflict because of competing and currently irreconcilable differences in the interpretation and meaning of that identity. The parade is thus, as Marston comments, an 'ethnic and national ritual embedded in a complex politics of domination and empowerment that has enabled the subversion of some identities and the assertion of others' (2002: 389). Space and place are central to this ritual. So too is the state, since it intervenes to adjudicate and regulate on the relationship between space (particularly public space) and identity.

The second example concerns research into resource conflicts. In 1995 Ismail Serageldin, then Vice-President of the World Bank, commented that 'if the wars of this century were fought over oil, the wars of the next century will be fought over water' (in Otis 2002). Recent events, in particular conflict in the Persian Gulf, have brought Serageldin's comments into sharp relief, suggesting that oil continues to be a source of resource conflict (Le Billon and El Khatib 2004). However, recent work by political geographers and others has started to foreground the importance of water as a source of conflict, as well as ways of managing and resolving such conflicts (see the recent special issue of *Political Geography*, edited by Sara McLaughlin Mitchell, 2006, on conflict and cooperation over international rivers). Conflict over water can occur for a variety of reasons. The first relates to the ownership of water, particularly whether it is seen as a common good or as private property. Concerns and conflicts over the privatization of water range from local to international (see Laurie 2007). The second relates to water as a scarce resource, where conflict arises over access to and use of water. The third relates to water and development, particularly through the building of large-scale dams that divert water and displace

Figure 19.3 Narmada Dam (Sardar Samovar Dam), India

people. These conflicts have been described by Vandana Shiva (2002) as either traditional or paradigm: the first are actual wars within communities, within countries, between countries and regions; while the second are conflicts over how we perceive and experience water.

Though not a political geographer, Arundhati Roy is often cited within geography, particularly in relation to her essays on the aftermath of 9/11, 2001, as well as on the conflict between India and Pakistan. One essay that has received little attention provides an excellent insight into the political geography of dam-building (Roy 1999). In it, Roy highlights the importance of dam-building to the Indian national project: in the years since independence, over 3,000 large dams have been built in the country. Roy estimates that up to 30 million people have been displaced by these projects, 60 per cent of whom are Adivasi (or Tribals), the poorest group in Indian society. One of the most recent large-scale projects is in the Narmada Valley (see Figure 19.3). The project consists of 30 large, 135 medium and 3,000 minor dams on the Narmada River and its tributaries. When completed, it is estimated that 350,000 hectares of forest and 200,000 hectares of cultivable land will be submerged, and up to one million people will be displaced. In addition to the displacement, it will affect the lives of 25 million people who live in the Narmada Valley. In her essay, Roy describes the various ways in which people affected by the project formed an action group aimed at preventing construction, held peaceful protests and went on hunger strike in an attempt to draw attention to the devastating effects of dam-building on people's livelihoods and on the environment. Roy's essay is one way of demonstrating conflict over water, a topic that is attracting growing attention in political geography.

KEY POINTS

- Conflict results from the pursuit of incompatible goals by different people, groups or institutions. Conflict can take a variety of forms, and may include violence.
- Political geographers generally focus on three main causes of conflict – territory, resources or ideology. In many cases, conflicts result from a combination of these three factors.
- Recent studies of conflict in political geography range from global conflicts, such as the 'war on terror', to more localized conflicts over public space in a city, to resource conflicts at variety of scales.

FURTHER READING

Flint, C. (ed.) (2005) *The Geography of War and Peace: From Death Camps to Diplomals*. Oxford: Oxford University Press.

Le Billon, P. (2007) 'Geographies of war: perspectives on "resource wars"', *Geography Compass*, 1(2): 163–182.

Marston, S.A. (2002) 'Making difference: conflict over Irish identity in the New York City St Patrick's Day Parade', *Political Geography*, 21: 373–392.

Roy, A. (1999) *The Cost of Living*. London: Flamingo.

20 POST-CONFLICT

Carl T. Dahlman

Definition: Life after War

A post-conflict period commences with the end of war. In common use, 'post-conflict' denotes the cessation of armed conflict and usually the start of a reconstruction and recovery period (Figure 20.1).

Figure 20.1 Destruction of German cities during World War II, like Nuremberg shown here after its capture by the US Army in 1945, was followed by an enormous post-war reconstruction effort and the creation of the modern European economy. (Courtesy of National Archives and Records Administration).

The term post-conflict emerged in the early twentieth century and is based on a number of underlying assumptions that held true during that period, but which are no longer fully applicable. This context frames current difficulties with using the term today. 'Post-conflict' implies, first, that formal states are the sole or at least primary party to a conflict and that they end wars on clearly negotiated terms. However, state on state wars, which some scholars call 'old wars', are not representative of most armed conflicts today, which makes it harder to identify the end of war and the beginning of something else except by the relative level of armed violence. Indeed, contemporary wars are increasingly fought by poorly defined forces under obscure or even shifting chains of command with transitory or non-existent relationships to formal states (Kaldor 1999). Formal treaty-making thus fails to address many of the issues 'on the ground' that fuel armed conflicts since the state parties to the treaty may have little practical control over the non-state parties that fought the war. A second assumption informing the term post-conflict concerns the role of combatants once a war is over. Since the term was developed when wars were fought primarily between states, the assumption was that soldiers either went home or demobilized after the fighting had ended in victory, defeat or a return to the status quo. Many armed conflicts today, however, do not end with clarity on the battlefield but instead produce an ambiguous and unstable political suspension of the war in which former combatants remain intermingled with each other and with civilians.

The question of who has power in a post-conflict society reminds us that reconstruction and recovery are executed by actors with varying capacities and specific visions of what those activities should accomplish. Thus the question of who funds and guides reconstruction may reanimate aspects of the conflict itself as old and new geopolitical interests clash over how to proceed or, more simply, who benefits (Caplan 2002; Fukuyama 2004). These problems remind us that post-conflict situations are often marked by the continuation of war by other means as, for example, competing partisan groups furthering their agendas during the redesign of constitutions, election systems and development aid (Foucault 2003: 15–19). If the term post-conflict is to be applied to the messy and contingent realities of life after war, then we must make a distinction based on the quality of the peace actually achieved. Peace studies provides some guidance with the notion of negative and positive peace. Negative peace describes a situation in which war is formally or practically concluded but the parties to the conflict continue to take other coercive measures towards particularistic goals or merely to retain gains made during the war. Positive peace, by contrast, is more than the end of violence and coercion; it seeks to reduce the structural violence that limits individuals and communities from enjoying political participation, economic opportunities, personal security and dignity, and rights and justice under the law (see Barash 2000: 61–64, 129–130).

Post-conflict periods are perhaps easier to identify with the passage of time, although more so in larger wars that produce clearer results than in civil wars and guerilla insurgencies. However, knowing when the post-conflict period starts can be just as uncertain as knowing when it is over. The beginning of the post-World War II period in Europe, for example, could easily be dated from a small number of events on the battlefield or at the negotiating table. From this moment, we begin counting the years of post-conflict recovery, but such a self-evident 'year zero' is less obvious in other wars. The war in Bosnia, for example, was concluded at the Dayton Peace Accords in November 1995 and, although the ceasefire held, the country remained divided by ethno-nationalists under a non-functioning government for years to follow. The end of the war in Vietnam should have meant peace for Cambodia, but instead a civil war led by the Khmer Rouge devastated the rest of the country, and this war was in turn put down by the Vietnamese in 1978. It was under a decade of de facto Vietnamese leadership, ironically, that Cambodian reconstruction and recovery unfolded during the 1980s (Gottesman 2004).

Identifying the end of a post-conflict period can be just as daunting – what constitutes closure, recovery, and a sense of having moved past the reasons for war? Fifty years after World War II, one author has put a definitive closing date on post-conflict Europe, by describing the entire Cold War period from 1945–1989 as merely 'an interim age' (Judt 2005). To wait such a long time for the resolution of ideological conflicts, however fundamental to Europe's course, risks losing sight of the more typical time horizon for the process of post-conflict reconstruction and reconciliation. In Japan, the six-year US occupation marked the formal post-conflict period although the miracle of its economic recovery took decades (Dower 2000). Twelve years after the formal end of the war in Bosnia, by contrast, the continuation of the international institutions established at Dayton points to the not-yet-complete effort to transform the country into a state ready for integration in NATO and the EU (Dahlman & Ó Tuathail 2005).

This uncertainty about when a post-conflict period draws to a close fixes our attention on the various tasks that occupy a polity after war.

Terms like 'reconstruction' and 'recovery' wrongly presuppose a polity needing no more than to fix damaged bridges and erect memorials commemorating the dead. In fact, post-conflict situations are more diverse and complex than this, but for convenience we can identify three areas of transformation: in political, economic and socio-cultural institutions.

Political Transformation

First and foremost, post-conflict periods are consumed with the renovation of political institutions. Where a recognized political entity remains in place with popular legitimacy the problems are fewer than where the war leaves no clear or commonly accepted authority. Despite local objections, Wilsonian ideals remain the norm for post-conflict societies, such as prescribing a course of elections, constitution-writing and legal reform, even where their prognosis for success is poor. Peace deals that end civil and transnational wars are beset by the reality that multiple parties and armies remain on the territory of the post-conflict state. Such deals usually seek to retain the sovereignty principle of territorial integrity but ignore the problem of effective sovereign control, which fades fast once outside Kabul, Sarajevo or Baghdad's Green Zone.

Economic Transformation

Economic activities comprise the second major activity of post-conflict periods, encompassing a wide range of market forces both domestic and international, licit and otherwise. Basic agricultural production is often the last productive activity abandoned during war and the first to return. Other sectors, especially construction and materials, electricity and utilities, and trade in basic consumables pick up quickly after war. For different reasons, public services and industrial production are slow to resume, the former because of weak government, the latter because of weak demand and access. Post-conflict economic initiatives face dramatically reorganized market realities, causing economic engines to stall in the face of domestic recession or because markets were captured by competitors during the war. Enterprises are often the bounty of war, too, sold off as parts for quick cash, destroying the ability to support wage labour or contribute to local economic expansion. More problematically, the illicit trade that developed during the war to profit from hungry populations and by arming combatants continues after war. These trade networks are often entwined with the political players who continue to profit from conditions of want. Profiteering is indeed hardly indistinguishable from the larger problem of corruption that plagues post-conflict politics.

Socio-cultural Transformation

War leaves deep social and psychological scars in a post-conflict society, effects that may remain unresolved long after the political and economic recovery is declared complete. These issues can be recognized across all social levels, giving shape to post-conflict life. In modern wars, national identities are often transformed into patriotic belief and profound personal investment in the mission. Disappointment is inevitable, however, even in victory, and the resulting soul-searching can produce crises in public confidence, a collapse of inter-communal trust, or else a desire for revenge and a search for the blameworthy (Schivelbusch 2004). In post-war Germany, for example, defeat led to an immediate and long-term reformation of what it meant to be German. In the post-bellum American South, by contrast, defeat was met with bitterness and a backlash of restrictive laws known colloquially as Jim Crow launched against freed slaves and their descendants. Commemoration and memorialization are functions of collective memory in public space, frequently drawing a sharp point as to who belongs where in a post-conflict society. The criminalization of war provides an uncertain and particular justice, but war crimes tribunals serve an important role by recasting great feats as shameful episodes, war heroes as outlaws.

Evolution and Debate: Post-conflict for Whom?

Post-conflict situations, then, are marked by an enormous number of tasks that remain sensitive to the terms of the original conflict. The issue becomes not what needs to be done, but how is it to be done and by whom? Two of the most successful post-conflict recoveries in the twentieth century were in Germany and Japan, but these were two formerly industrial countries, one formerly democratic, both with relatively homogeneous national cultures. Both were occupied after a decisive defeat, their humiliation contributing to a political quiescence that turned reform and reconstruction into a kind of national penance. Nation-building was achieved with a good deal of US involvement and investment, as was the reconstruction of Western European allies through the Marshall Plan. Although the Soviets vetoed western assistance for the reconstruction of Eastern Europe, the Soviet Union's own recovery was achieved rapidly. This was possible because a newly patriotic and multiethnic civic identity, buoyed by victory over Germany, fully restored pre-war Soviet rule and economic policies. The Soviets were also able to take advantage of an enormous resource base to fuel rebuilding (Hosking 2001: 509–526). Most post-conflict societies, however, have relied on foreign assistance, expertise and guidance, often through occupation, thus returning our attention to the goals of outside powers.

Foreign intervention in post-conflict societies raises three specific dilemmas. First, the need to ensure security in post-conflict societies, especially after civil wars, often falls to interventionist forces, normally under the auspices of the United Nations or designated collective security organs like NATO. These missions are meant to either impose a cessation of hostilities – peace enforcement – or else to ensure that a ceasefire lasts – peacekeeping – thus giving political solutions time to gel. Occupying or conquering armies may also inherit peacekeeping duties once the enemy forces are defeated. Despite the importance of peacekeepers, especially for weakened states, the presence of foreign soldiers often puts post-conflict states in a weaker position regarding their own affairs. This reliance on foreign forces presents a legitimacy problem for post-conflict governments, and domestic pressures in countries sending peacekeepers often puts a short timeline on deployment. Furthermore, the role and right balance of military to civilian interventions, their responsibilities, and their duration are always a matter of debate for the parties involved. Such deliberations are unlikely to produce much agreement, however, since each intervening power and each post-conflict society has different constraints and expectations of what should be accomplished.

A second, and perhaps more contentious, issue surrounds the question of nation-building. The restoration and reform of civilian rule in post-conflict societies is often given much closer scrutiny than either military or economic issues, since nation-builders consider responsive political institutions the quintessential means of avoiding further war (Fukuyama 2004; Ignatieff 2003). For this reason, it is more accurate to call these activities state-building. The term 'state-building' is also more applicable since many contemporary post-conflict societies involve competing nationalisms and other political identities. The dominant vision of the US and European experience in state-building assumes that democratic elections, the rule of law and free markets can all be built in the wake of war (Fukuyama 2004). Many contemporary post-conflict societies, however, remain internally riven by the factions, identities and competing interests that caused the war. These same legacies of the war stymie the third, economic, goal of international interventions. Rather than starting from an economic blank slate, the wealthy trading interests created by wartime deprivation and profiteering seek to maintain their control over markets. This means that external economic intervention programmes have limited effect because they cannot exclude the existing powerful actors who have either already captured the valuable means of production, distribution and retail or who are in a position to take over new ones.

Case Study: Afghanistan

Afghanistan serves as an unfortunate example of the complex geopolitical and humanitarian issues that arise in post-conflict settings (Figure 20.2). The country has been devastated by three decades of war that have left most of its 32 million citizens among the poorest in the world. The Soviets invaded in 1979 to support the central government, leading to a decade of fighting with the anti-communist mujahideen. After the Soviets withdrew and the government collapsed, the mujahideen began fighting other militias formed under local tribal authorities or regional ethnic leadership. The Taliban, an Islamist militia run by fundamentalist Islamists and backed by Pakistan, attracted many of the disaffected and uprooted young men from the slums and refugee camps in and around Afghanistan. They quickly took over large parts of the country and brought relative calm to people abused by the rival factions. But the Taliban imposed a strict interpretation of Koranic law that sought to remove 'western' influences and install in their place repressive strictures most notably, but not only, on women and gender relations. They took Kabul in 1996 and extended their control over almost all the country by 2000, after which they began to curtail the production of opium that had fuelled the war economy, including the Taliban's own rise to power. The relationship between the Taliban and Osama bin Laden, who had training camps in Afghanistan, was the decisive factor in the US decision to invade the country and overthrow the Taliban as a government that sponsored the terrorists responsible for the attacks of September 2001.

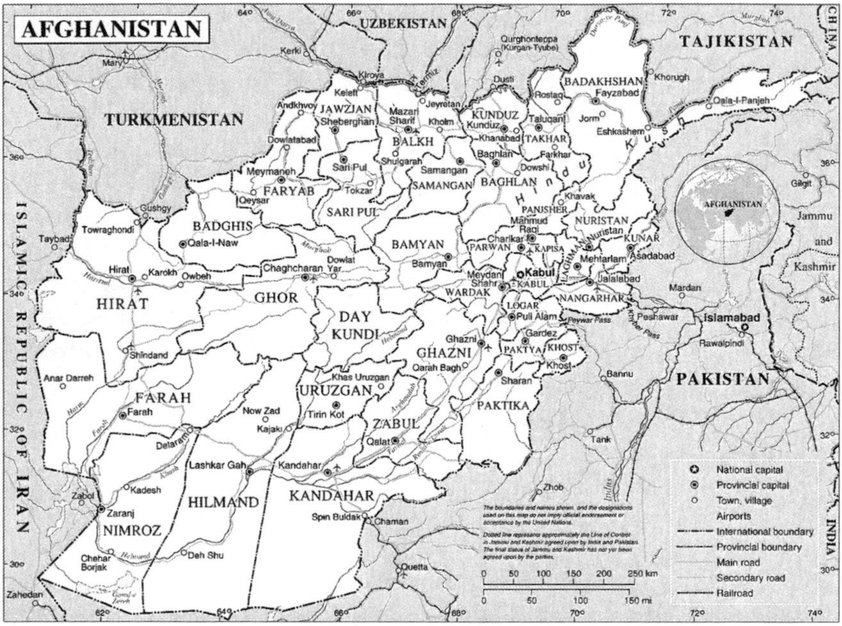

Figure 20.2 Map of Afghanistan based on UN map 3958, Oct. 2005.

The United States, along with local forces, began their campaign against the Taliban in October 2001 and took Kabul in November. The Taliban was pushed into the mountains along the Afghan–Pakistan border, and it seemed the country was moving into a post-conflict period. However, the vacuum left by the departure of the Taliban meant that some leaders tried to re-establish areas of tribal authority that had been recognized by previous governments or else to carve out new, sometimes ethnically defined, regions of their own. Afghanistan's sovereignty was, therefore, contested at the most fundamental level of political authority, delaying all post-conflict activities. A UN-sponsored negotiation among Afghani leaders and western diplomats in Bonn, Germany, in December 2001 produced a plan for establishing a new government. The first step was the creation of an Afghan Interim Authority, chaired by Hamid Karzai, followed by a *loya jirga* (grand council) in June 2002 that named Karzai President of the Transitional Administration.[1] A second *loya jirga* produced a constitution in January 2004, under which Karzai was elected President later that

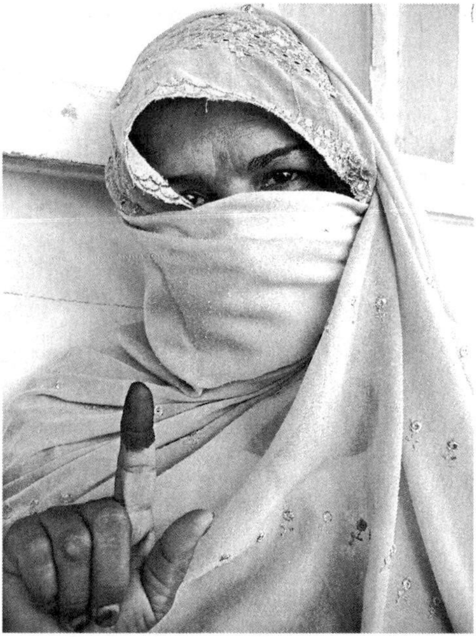

Figure 20.3 Purple ink to prevent vote fraud marks the finger of an Afghan women who cast a ballot in the 2005 parliamentary elections. (Photo by Staff Sgt. Jacob Caldwell, courtesy of the US Army).

year, followed by the election of a national assembly in 2005 (Figure 20.3). These elections marked the formal end of the Bonn Process, but the sovereign territory accorded to the new government of Afghanistan did not coincide with that under its control.

During the post-Taliban period, the reach of the governing authorities in Kabul has not extended far beyond the capital, at least not without foreign military assistance and extensive deal-making with regional warlords. Local militias, not all of whom support a strong central government, have provided de facto security across much of the country since the fall of the Taliban while the US has concentrated on securing Kabul and fighting the remnants of the armed Taliban forces in the countryside. In December 2001 the UN Security Council authorized the International Security Assistance Force (ISAF) to protect Kabul while the Afghan Interim Authority and United Nations got under way there. The Kabul mission was backed by troops from NATO countries and aided by several other countries, while the US took responsibility for security in the rest of the country, and that was largely limited to operations in pursuit of the Taliban. In 2003, NATO took formal command of the ISAF mission and the UN Security Council expanded the ISAF mission beyond Kabul as training of the new Afghan National Army continued. These military deployments also included provincial reconstruction teams whose primary task was to provide security in response to local needs, while secondarily supporting humanitarian assistance and small public works projects (ICG 2007). The list of post-conflict tasks in Afghanistan was quite long and many of these were indistinguishable from development needs that preceded the Soviet invasion: transport and utility infrastructure, basic social services and economic diversification.

US-led state-building in Afghanistan has created centralized political institutions in a country with a weak tradition of such authority, hence the need for the neo-traditional *loya jirga* and Karzai's subsequent horse-trading with regional warlords, especially among the various ethnic groups, to try and achieve domestic authority (Rubin 2002: 22–32). As with many post-conflict interventions, the presence of US, NATO and UN staff, international NGO workers and the delivery of foreign aid in Afghanistan's economy caused it to expand rapidly as the infusion of cash lubricated the return of local markets. Nevertheless, Afghanistan remains a poor country with an infrastructure far too inadequate to attract foreign investment even if the security conditions improved. The re-emergence of opium poppy production as the country's leading cash crop has meant that the most widely available means for local economic recovery has itself become a target of the international forces. The United States has worked to destroy poppy crops but at the expense of alienating local communities from the larger nation-building effort as well as from the government in Kabul (Anderson 2007).

Six years and billions of dollars later, the US, its allies and international aid agencies have at best produced an unstable and weakly democratic government, with few signs that state-building or economic recovery extends far from Kabul. The 80-odd political parties recognized by the Afghanistan government are, unfortunately, more a testament to competing provincial and tribal interests than evidence of a flowering pluralistic democracy. The creation of public roles for women and the reopening of schools for girls in some areas have not led to the removal of the *chadri* by most Afghan women, as naively hoped by some – most of Afghan society remains socially conservative.[2] Afghanistan remains plagued by landmines and unexploded ordnance, left by the various armies that fought there in recent decades (ICBL 2006). Afghanistan's economy remains small, amounting to $800 per capita GDP, although this does not figure in the revenues from poppy production, and other indicators suggest that it remains near the bottom of the UN human development index. A vision for continued state-building, the Afghanistan Compact, was adopted in London in 2006 to replace the Bonn Process. The Compact pledged enormous international assistance for security, state-building and economic development (ICG 2007). Progress on the Compact has been slow, however, primarily because of the renewed Taliban insurgency, fuelled in part by revenues collected from opium smugglers (Anderson 2007). Despite the significant military, political and economic interventions, Afghanistan remains a fragmented geopolitical space. Central government authority competes with regional and local bases of power, forming localized zones of sovereign authority, traditional rule or rebellious exception. Battlegrounds with ongoing armed conflict contrast with areas of relative stability and development. Afghanistan shows us that the transformative process of moving beyond conflict is not linear but rather reveals the situated geographies of authority and violence that keep a country somewhere between war and peace.

KEY POINTS

- Post-conflict describes the condition that comes after armed conflict but not necessarily the resolution of that conflict. In some ways, the post-conflict period is the continuation of war by other means.
- Post-conflict periods are not always easy to identify, nor are they necessarily a durable condition.
- Post-conflict spaces are shaped not only by the terms that ended the war but often by the agendas and capacities of third parties who assist in reconstruction and state-building efforts.

NOTES

1 The *loya jirga* was adopted because it represented a traditional means of decision-making by a collection of tribal elders and traditional leaders from across the cultural groups that comprise

Afghanistan. Although hardly a democratic institution, it was a means of extending legitimacy to a new central government from across the decentralized bases of Afganistan's *de facto* ruling authorities.

2 The *chadri* or *burqa* is the traditional full body covering worn by many women in Afghanistan and was mandatory under the Taliban.

FURTHER READING

Barash, D. (2000) *Approaches to Peace: A Reader in Peace Studies*. New York: Oxford University Press.

Caplan, R. (2002) *A New Trusteeship? The International Administration of War-torn Territories*. New York: Oxford University Press.

Fukuyama, F. (2004) *State-building: Governance and World Order in the 21st Century*. Ithaca, NY: Cornell University Press.

Schivelbusch, W. (2004) *The Culture of Defeat: On National Trauma, Mourning, and Recovery*. New York: Metropolitan Books.

21 TERRORISM

Carolyn Gallaher

Definition: Using Violence to Secure an Objective

Terrorism is a tactic of war designed to intimidate, bully or otherwise frighten a population, usually civilian, by using or threatening to use violence. Terrorist attacks are relatively rare. Some studies suggest that a person is more likely to be struck by lightning than to fall victim to a terrorist attack (Mueller 2007). Terrorism is an effective tactic, however, because most terrorist attacks have 'a random quality' (Kliot and Charney 2006; Mueller 2007). Attacks are generally carried out without prior warning and in everyday places, such as marketplaces, churches, offices or bus stops. Some victims are killed or maimed by explosive devices. Others are rounded up and raped in front of family members. Many see their homes ransacked and burned to the ground.

While people from any demographic group may be victimized by a terrorist attack, studies suggest that most attacks are not as random as they appear. Vulnerable groups in societies are often at greater risk of being victims of a terrorist attack. In places where terrorists target public transportation, for example, poor people are at a greater risk than the rich since they rely more frequently on public transportation to get around. Likewise, young people are more likely to be victimized by a terror attack than older cohorts because they tend to gather in groups at public places such as restaurants, clubs and cafés (Canetti-Nisim et al. 2006).

Because terrorism is a tactic of war, any agent involved in war may use it, including states, irregular (i.e. non-state) forces and even individuals. However, the term is most often associated with non-state actors, such as guerillas, paramilitaries and insurgents (Nagengast 1994). The modern nation-state system was based on states monopolizing the means of violence within their respective borders. They were able to legitimize these monopolies by promising to protect citizens from outside attack. Rhetorically, states have guarded these monopolies by labeling violence by non-state entities as 'terrorist' and 'illegitimate' (Nagengast 1994). The term tends, therefore, to be associated in common parlance with non-state actors and so when scholars talk about terrorism directed by state armies, they usually signify as much by using the term 'state terrorism' (see Wright's 2007 study on state terrorism in Latin America).

Terrorism is used to secure a political objective (Crenshaw 1981; Whittaker 2001). Political objectives may stem from ideology, religious conviction or world view. Terrorists use terrorism to gain power or force others to behave in a certain way (US Army 1990; Whittaker 2001). When Taliban fighters attacked girls' schools in Afghanistan, for example, they hoped to discourage women from participating in the public life of the state. Likewise, when the Irish Republican Army (IRA) bombed locations in London in the 1990s, it hoped to convince British civilians to rally the state to pull out of Northern Ireland. In a similar fashion, the Russian Army attacked civilians during the Chechen War because it hoped to root out combatants living in their midst. Terrorism is distinct both from individual acts of violence, such as murder, and from group violence driven by personal goals, such as criminal enrichment (US Army 1990; Whittaker 2001).[1]

The Changing Face of Terrorism

After World War II the use of terrorism as a tactic of war increased. In particular, its use by non-state actors is on the rise. By the 1950s and early 1960s, guerilla groups in places as diverse as Northern Ireland, Algeria and Kenya (among other places), employed terrorism in an effort to overthrow colonial powers, leading many people to associate terrorism with revolutionary movements to end colonialism (Kliot and Charney 2006). It is important to note, however, that guerilla warfare and terrorism are not interchangeable terms (Hoffman 1998). Some guerilla groups opposed using terrorist tactics, except in rare cases. In *Guerilla Warfare*, for example, Che Guevara (see Figure 21.1) argued:

> It is necessary to distinguish clearly between sabotage, a revolutionary and highly effective method of warfare, and terrorism, a measure that is generally ineffective and indiscriminate in its results, since it often makes victims of innocent people and destroys a large number of lives that would be valuable to the revolution. Terrorism should be considered a valuable tactic when it is used to put to death some noted leader of the oppressing forces well known for his cruelty, his efficiency in repression, or other quality that makes his elimination useful. But the killing of persons of small importance is never advisable, since it brings on an increase of reprisals, including deaths. (1998: 22)

Figure 21.1 A Cuban stamp commemorating Che Guevara

Separatist and nationalist groups have also used terrorism in the decades since World War II. Many of these groups called for separation or autonomy for an ethnic and/or religious group. The ETA (a Spanish language acronym for Basque Homeland and Freedom), a guerilla group in northern Spain, for example, has been fighting the Spanish government since 1959 for an independent Basque homeland. Likewise, Chechen rebels fought two wars with the Russian state in the 1990s in an effort to create an independent Chechen state. Both groups used attacks against civilian populations. However, while the Spanish government's response to ETA has tended to be muted, the Russian Army employed a razed earth policy in Chechnya that resulted in thousands of civilian deaths (Tishkov 2004).

Terrorism has also increased in so-called failed states, where a variety of groups compete to monopolize the means of violence and the control of political and economic structures it entails. During the course of the recent war in the Democratic Republic of the Congo (1997–2003) there were as many as 18 identifiable non-state groups and six state

Figure 21.2 Propaganda poster with image of Osama bin Laden on it. (Courtesy of the Defense Department - Defense Visual Information Center)

armies fighting one another in the country's vast eastern half. Most of these groups used classic terror tactics, including civilian massacres, rape and burning villages to the ground. Experts estimate that nearly four million civilians died during the war (Amnesty International 2003).

A relatively new breed of terrorism – transnational terrorism – has also appeared in the last twenty years. Transnational terrorist groups are not based in a single country and claim to represent individuals spread across national boundaries. The group Al Qaeda, for example, was founded by a Saudi national, Osama bin Laden (see Figure 21.2), and the group's second-in-command is from Egypt. The group also has (or had) bases of operation in several countries, including Sudan, Afghanistan and Pakistan. Individual operatives of the group are active in at least a dozen more states, including those in the western industrialized world.

Evolution and Debate: Do We Study Terrorism or Combat It?

The discipline of Geography has not usually been a source discipline for studies of political violence, generally, or terrorism more specifically (for exceptions see Flint 2003b; Gallaher 2004; Pickles 2000). Many geographic studies of violence and terrorism do not employ an explicit theoretical framework. Geographers who do frame their work theoretically tend to situate their analyses in one of two broad theoretical frameworks: political economy or geopolitics. On occasion, authors borrow theoretical insights from both approaches or rely on debates specific to subdisciplines within the field.

Political Economy Approaches

The political economy approach examines the factors that contribute to violence more than the factors related to the choice of specific tactics, such as terrorism. Neo-Malthusian approaches, for example, argue that environmental scarcity (brought on by natural factors like drought or human factors like resource depletion) contributes to war and to tactics like terrorism (Homer-Dixon 2001; Kalpan 1994). The most ardent proponent of this view is Thomas Homer-Dixon. Although he is a political scientist by training, Homer Dixon's work (2001) addresses a classic geographic research area, the nature/society nexus. Other geographers, like Nancy Peluso and Michael Watts (2001) argue that environmental scarcity is often an expression of capitalist relations and the limitations to access that it brings. As such, forms of governmentality, which often limit access, cause more violence than scarcity itself. Likewise, Le Billon (2004) notes that war zones in Sierra Leone and Liberia, which neo-Malthusians blamed on resource scarcity (see Kaplan 1994), actually have an abundance of resources. Their problem is not scarcity, but the inability to translate 'resource exploitation into political stability and economic development' (1994: 22).

Other political economic approaches to violence and terrorism in geography examine the relationship between modernity (and its attendant emphases on 'development,' creative destruction and democratization) and the production of ethnic difference, upon which so many contemporary conflicts turn. In this configuration, violence is not an 'interruption' or obstacle on the path of progress; it is part and parcel of modernization (Pickles 2000). Scholars argue that the scale of violence in the twentieth century – to date the most violent century on record – is the result of the creation of the modern state and standing armies, and its monopoly of the means of violence (Mann 1999; Pickles 2000). States set about marking territory and claiming sovereignty over it and, once this was accomplished, created national myths about who a 'true' citizen was and by extension who he/she was not. Those cast outside these nets are deemed ethnically, religiously, or nationally different and are marginalized, even repressed, by the modern state. Likewise, the violence associated with groups calling for territorial autonomy – for ethnic, religious or national groups – is connected to the modernization project. They represent a response to systematic marginalization in the state system and an attempt to create alternative spaces free of it.

Geopolitical Approaches

The geopolitical approach to terrorism analyses how states define terrorism and how these definitions shape state responses to it. Scholars using this approach note that states employ geopolitical narratives to make sense of threats and to delineate security responses to them. While traditional scholars (in and out of geography) view the reason for threats as self-evident, critical geopoliticians argue that threats are the products of state narratives and that different narratives can produce different threats. Several political geographers have noted, for example, that the rigid narratives employed by the US during the Cold War to define enemies of the West created 'threats' that posed little real danger to Western interests. US involvement in Central American civil wars during the 1980s, for example, was based on the assumption that central American guerilla campaigns were part of a global communist project aimed at world domination, when a more nuanced analysis demonstrates that most of the region's civil wars were triggered by local dynamics (Ó Tuathail 1986). Many were an attempt to throw off the vestiges of colonial social relations that the region's *caudillo* governments had fostered since independence. Labeling guerilla groups in El Salvador and Nicaragua as terrorists, however, permitted the American foreign policy establishment to ignore the very social and economic injustices that had made communism an attractive ideology in the first place.

More recent analyses have focused on the post-Cold War era. Several scholars have unpacked the geopolitical narratives underlying the so-called 'War on Terror'. In *The Colonial Present* (2004) Derek Gregory argues that American President George W. Bush's 'War on Terror' relies on a false dichotomy – civilized/uncivilized – that allows the US foreign policy establishment to conflate different, even warring parties into an undifferentiated enemy. Ignoring differences between Al Qaeda, Saddam Hussein's Iraq, and the Taliban in Afghanistan permitted the US to pursue a foreign policy built on a crusader ethos.

Empirical Studies

Other geographic studies of terrorism are empirically focused rather than theoretical explanations that situate terrorism in wider contexts. A number of geographers approach the study of terrorism by asking classic location analysis questions (Kent 1993; Kliot and Charney 2006; McColl 1969). Some studies note that anti-state groups base their operations in rural or otherwise difficult to access areas (McColl 1969). A number of terrorist groups are successful at exploiting local knowledge of the environment to recruit new members and thwart counter-insurgency efforts (Kent 1993). Other studies note that attacks are concentrated in urban areas with the greatest population densities. The implication is that terrorists choose their targets in a rational manner – seeking locations that allow for the greatest possible death tolls (Kliot and Charney 2006).

Other geographers have approached terrorism as a hazard (Galloway 2003; Mitchell, J. 2003). In this approach spatial vulnerabilities to attack are identified and mechanisms for decreasing vulnerabilities are considered. Many scholars using this approach argue that Geographic Information Systems should be employed to manage the hazard of terrorism and that geographers should market their unique contribution to terrorism analysis (GIS and location analysis) to government and private sector groups interested in preventing terrorism. It is argued that doing so would elevate geography's profile and contribute to an ongoing resurgence of the discipline in secondary schools and colleges and universities.

A Discipline Considers its Role in the 'War on Terror'

After the terrorist attacks in the United States on 9/11, 2001, academics sought to make sense of the tragedy and to proffer suggestions for preventing future attacks. In January 2002 the Association of American Geographers organized a terrorism and geography workshop in Washington, DC. The workshop eventually led to a book, *The Geographical Dimensions of Terrorism* (2003), a collection of essays edited by Susan Cutter, Douglas Richardson and Thomas Wilbanks. The book outlined four issue areas to which geographers could contribute knowledge: reducing vulnerabilities to threats, detecting threats, reducing threats, and improving responses to terrorism. The book also posited a very specific policy goal: the creation of a GeoSecurity Information Office within the Department of Homeland Security (Wood 2003).

Although most geographers lauded the group's efforts to address such a pressing issue and to do so in a timely manner, the book was criticized for defining geography's contribution in a narrow, technocratic way (DeBlij 2004; Griffith 2004). Harm DeBlij, who described the book as 'turgid and uninspiring', identified the lacuna made possible by its technocratic focus:

> Nothing is said here about cultural or physical environments in source areas where terrorist plots originate, very little is discussed about the flow of resources (diamonds, drugs, oil, mosques, charities) that sustain the campaigns, and no mention is made of the diffusion processes that continue to propel the widening circle that eventually touched the American 'homeland'. (2004: 994)

DeBlij also chastised the book for failing to consult regional specialists in the discipline who could help situate the wider context of Islamic-based insurgency.

The debate about whether Geography should define its role in technocratic terms or stretch more deeply into the discipline's historic focuses (including regionalism and political economy) has occurred in other wings of the discipline as well. A 2003 article in the *Professional Geographer* by R.A. Beck, and the response it generated, provides another case in point. Beck's article, which was about the use of remote sensing and GIS as a counter-terrorism tool in Afghanistan in coordination with the US government, drew condemnation from John O'Loughlin (2005a), then editor of the journal *Political Geography*. In particular, O'Loughlin objected to the study's decision to comply with government restrictions regarding the timing of the publication and the omission of a complete methodology review: Beck delayed publication of the article and omitted full information in order to protect the US military effort there. O'Loughlin argued that the publication of Beck's article broke professional and ethical norms:

> Since when do AAG journals engage in this kind of obvious self-censorship? How can a study be replicated (the basis of scientific advancement) if some of the methods and information are withheld? Has the Association inadvertently or surreptitiously taken sides in the 'war on terrorism' without the knowledge of the members through violation of a basic tenet of academic review and publication? (2005a: 589)

Debates like these are likely to continue not only because war tends to spark debate, but because the discipline's long-standing fault lines – between technocrats and theoreticians, generalists and regionalists, and mainstream and critical approaches – continue to rumble beneath the surface.

Case Study: Suicide Terrorism in Israel

Suicide terrorism is a type of terrorism where the perpetrator commits suicide in the course of carrying out an attack on others. Many suicide terrorists strap bombs to their bodies and detonate them in places where crowds gather, such as markets, hotels, and buses. Other suicide terrorists use machines to launch their attacks; Al Qaeda hijacked commercial airplanes and flew them into the World Trade Center in New York City and the Pentagon in Washington, DC.

Suicide missions in the course of war have been a feature of many known wars. During World War II, for example, the Japanese employed kamikaze pilots to dive-bomb planes into Allied targets in the Pacific (Axell and Kase 2002). The Tamil Tigers have periodically used cyanide suicide attacks in their battle against the Sri Lankan government (Laqueur 1999). Palestinian groups have also used suicide attacks, particularly during the first and second Intifada (Kliot and Charney 2006).

Interpretations vary as to why Palestinian armed groups employ suicide as a terror tactic. Some scholars argue that Palestinian groups such as Islamic Jihad and Hamas use suicide bombings because they lack access to conventional weaponry and tactics available to the Israeli Defence Forces: Hamas is fighting an asymmetric war and using the tactics available to the weaker side (Wyne 2005). Others argue that suicide terrorism resonates in cultures where martyrdom has a central place in religious belief systems (Laqueur 1999).

In a recent study of suicide terrorism in Israel, the Gaza Strip and the West Bank between 1994 and 2005, geographers Nurit Kliot and

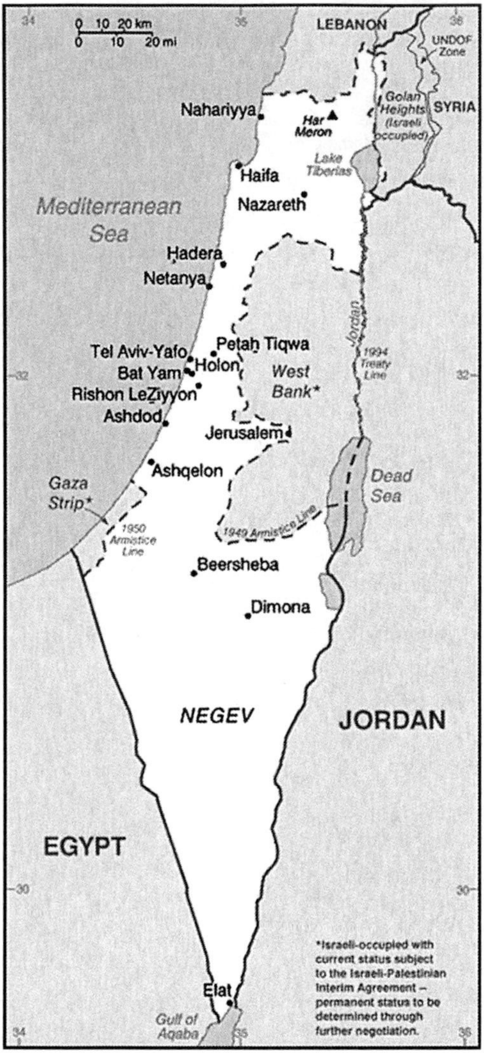

Figure 21.3 Map of Israel with major cities. (Courtesy of the University of Texas Perry-Castañeda Library Map Collection)

Igal Charney (2006) found several spatial patterns associated with suicide terrorism. First, the majority of attacks occurred in Israel's largest cities (see Figure 21.3). Jerusalem recorded the largest number of attacks, followed by Tel Aviv, Haifa and Netanya (see Figure 21.3). The

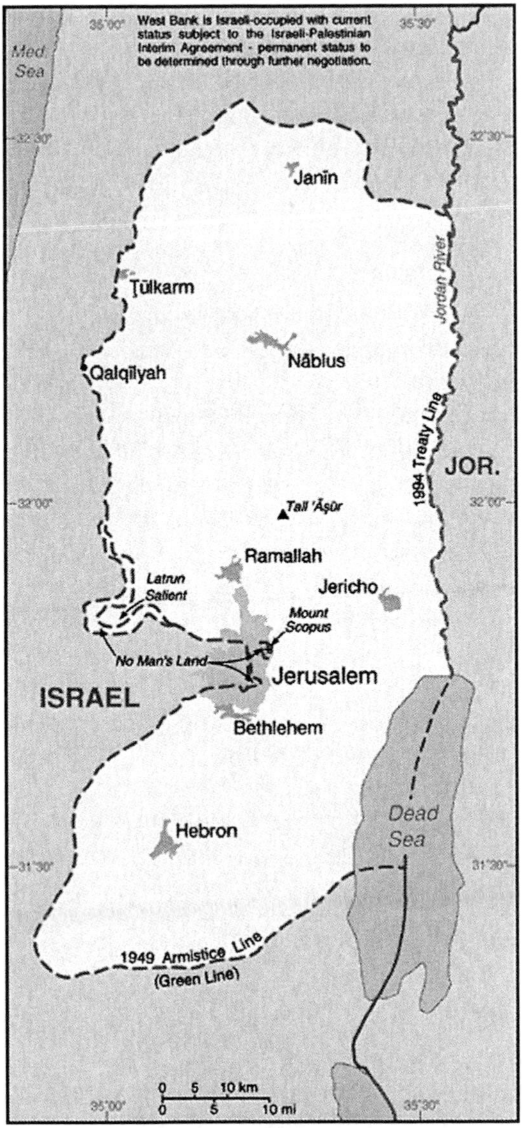

Figure 21.4 Map of the West Bank with major cities. (Courtesy of the University of Texas Perry-Castañeda Library Map Collection)

authors note that this finding 'confirm[s] the role of agglomeration in target selection'. Second, the authors note that terrorists select locations based on the relative ease of movement into them. The top four cities for suicide missions are 'highly accessible by car and transit as they serve as national and regional hubs' (2006: 360). These patterns demonstrate a clear friction of distance function at work in the site selection of terror attacks i.e. the further away a location is in distance or time travel, the less likely it is to be targeted for attack. The authors do note, however, that the Gaza Strip and Jewish settlements have fewer attacks than might be expected. They speculate that the lockdown of the Gaza Strip prevents a higher number of attacks there than would otherwise be the case given its density. Likewise, Jewish settlements, which are highly controversial for Palestinians, tend to have especially high levels of security.

Kliot and Charney (2006) also tried to identify whether certain cities 'produce' more terrorists than others (as measured by place of origin of the terrorists in their data sets). They found no discernible pattern, but they did find a geographic concentration in the location where attacks are planned. Most attacks originated by Islamic Jihad were planned in Jenin while those organized by Hamas were planned in and around Nablas (see Figure 21.4). The number of attacks was found to decrease as distance from these planning centres increases. Of the 118 attacks in their database, 54 were launched within 30 kilometres of the city in which they were planned, 28 were launched between 31 and 60 kilometres, and 10 were launched between 61 and 90 kilometres. Moreover, while 26 attacks occurred 90+ kilometres from the site of planning, Kliot and Charney note that most of these attacks were located in big cities. In short, suicide terror attacks follow a classic distance decay pattern. Unless an attacker was going for a high-value target (i.e. a big city), he/she was most likely to stay close to 'home'.

KEY POINTS

- Terrorism is a tactic of war designed to intimidate, bully and otherwise frighten a population, usually civilian, by using or threatening to use violence.
- Terrorism is a tactic of war, so any party involved in war (states, guerilla groups, paramilitaries, etc.) may use it.
- The term 'terrorism' is most commonly associated with non-state actors, so when scholars discuss terrorist tactics used by states they use the term 'state terrorism'.

- Geographers use two frames to examine terrorism – political economy and geopolitics. Those who use a political economy framework examine the structural factors that contribute to non-state violence. Those who use a geopolitics framework examine how states define terrorism and how these definitions shape state responses to it.
- Terrorist incidents are not randomly distributed in space. Incidents of terrorism tend to follow clear spatial patterns, including distance decay.

NOTE

1 Violence by criminal syndicates is not usually described as terroristic. However, if a criminal syndicate were to bomb a government building to discourage the prosecution of a crime boss, for example, it would be considered a terrorist operation

FURTHER READING

Flint, C. (2003b) 'Terrorism and counterterrorism: geographic research questions and agendas', *The Professional Geographer*, 55(2): 161–169.

Kliot, N. and Charney, I. (2006) 'The geography of suicide terrorism in Israel', *GeoJournal*, 66: 353–373.

Le Billon, P. (2004) 'The geopolitical economy of "resource wars"', *Geopolitics*, 9(1): 1–28.

Nagengast, C. (1994) 'Violence, terror, and the crisis of the state', *Annual Review of Anthropology*, 23: 109–136.

22 ANTI-STATISM

Carolyn Gallaher

Definition: Opposition from Above and Below

Anti-statism is a term used to describe opposition to the state and its power to regulate social, economic and political life. Anti-statism can take the form of policies aimed at hollowing out the state, or as a form of protest or social movement against the state. Neoliberalism is an example of the former. (For a complete discussion of neoliberalism see Harvey 2007 and Chapter 13.) The militia movement in the United States is an example of the latter (Gallaher 2003). As these examples indicate, anti-statism can emerge from 'above' or 'below'. Neoliberal policies are most often designed and implemented by states or international organizations, while social protest against the state often emerges on the street, at the local level.

The state that anti-statism opposes (albeit to varying degrees) emerged out of the Peace of Westphalia, a set of treaties in 1648 that ended the Thirty Years War in Europe. After 1648 states became the dominant form of political organization, replacing kingdoms and empires and diminishing the power of so-called sovereign rulers. The rights and obligations of states were defined by their boundaries; states could control what happened within them, but were prohibited from interfering with events outside them except in matters of defence.

The emergence of the modern state system was facilitated by Europe's colonial ventures and the rise of merchant capitalism that accompanied them. The new markets that built up around colonial loot provided an economic base outside standing feudal arrangements that states could tax. States used the proceeds to create standing armies, monopolize the means of violence and build state bureaucracies. Not surprisingly, the two systems – one political, the other economic – soon became inextricably linked, with the health of the market and state seen as one and the same.

Contemporary anti-statism, whether from above or below, is often a critique of the bureaucratic nature of the state. In anti-statist discourse the state is often presented as a juggermaut that squashes human liberty. Milton Friedman (1962) equated government regulation with slavery while John Trochman, co-founder of the Militia of Montana, described Federal Bureau of Investigation (FBI) agents as 'tyrants' and 'jack-booted thugs' (Gallaher 2003).

Evolution and Debate: Opposition of Varied Stripes

While the modern state system has always had its detractors, formal, intellectual opposition only emerged during the late nineteenth and early twentieth centuries. Though they varied in approach, opponents generally agreed that the state system worked against humans' natural tendency to meet their needs in relatively small social groups, such as the village, church or trade union (Laborde 2000). They argued that scaling up social organization to the level of the state allowed state imperatives to trump those of more organic groups. They also believed that a primary imperative of the modern state was territorial expansion, which set the stage for violence between states to become routine.

Marxists were some of the earliest anti-statists. Indeed, the idea that the state system was not organic complemented the Marxist view that state-driven capitalism did not operate in the best interests of workers. However, most Marxists did not reject the state system per se, so much as its capitalist form. Marx and Engels argued in the *Communist Manifesto* (1848) that the revolution would begin by workers taking over state structures. Marxists attracted to the anti-statist message eventually developed what they called an anarchist position; workers' associations, based in factories or neighbourhoods, should form the basis of governance rather than states. (For more information on governance see Chapter 3.)

Pyotr Kropotkin and Élisée Reclus: Geography's Anti-statist Legacy

Although anarchism has never had a huge following in the discipline, Richard Peet (1998) notes that geography and anarchism have had

Figure 22.1 Pyotr Kropotkin

'more than casual relations'. Peet posits that the discipline's historic focus on blending the physical and the human complements anarchist views that naturalize human behaviour and its spatial expression. Pyotr Kropotkin and Élisée Reclus are the discipline's most prominent anarchists.

Kropotkin, born in 1842, is perhaps the better known of the two (see Figure 22.1). The son of a Russian prince, Kropotkin renounced the monarchy in his twenties and affiliated with a number of burgeoning socialist associations across Europe. He was arrested in both Russia and France, and kicked out of Switzerland for allegedly seditious behaviour. In 1886 he settled in London. He eventually returned to Russia in 1917 after the Russian Revolution.

Kropotkin's work was devoted to theorizing the spatial elements of anarchism. His most famous essay, 'What Geography ought to be' was published in 1885. Kropotkin believed that human societies are inherently cooperative and that mutual aid was the basis of human survival. He did not deny competitive urges in human groups, but he thought they were sporadic emotions rather than the motor of human growth (as capitalists claimed). Kropotkin theorized that the cooperative spirit could only function properly under decentralized forms of governance. In this manner he rejected the scale of the modern nation-state and resisted the idea, then percolating through the discipline of Geography, that states had an organic need for territorial expansion (e.g. Friedrich Ratzel's notion of *lebensraum*). Kropotkin was equally suspicious of capitalism, which he viewed as destructive of the moral basis of human society and antithetical to its 'natural' geographic expression. It is worth noting, however, that when Kropotkin returned to Russia he is said to have been appalled by the authoritarian version of communism instituted by the Bolsheviks and the territorial imperative they claimed on its behalf (Peet 1998).

Élisée Reclus was less well known in geography than Kropotkin, but he maintained more formal ties with the discipline. He was a professor of comparative geography in Brussels and wrote widely in the field. Like Kropotkin, Reclus was interested in the social applications of Darwin to human geography and sought to develop a notion of social Darwinism that put mutuality rather than competition at the centre of human progress. Reclus believed that territorial boundaries were artificial, and did not correspond with the 'natural' regions of humankind. He rejected the discipline's focus on the geopolitical imperatives of states, arguing that geography would do better to study class struggle, individual will and the search for equilibrium between the two in the natural environment (Fabrizio 2007; Preface to Reclus' *L'Homme et la terre*, 1991). Indeed, Reclus saw violence and disorder as stemming from the state's territorial imperative, which he believed put humankind in opposition with nature.

Contemporary Forms of Anti-statism

While the intellectual lineage of Kropotkin, and to a lesser degree Reclus, is well acknowledged in histories of the discipline (Livingstone 1992; Peet 1998), few contemporary geographers have taken up the analytic or political mantle of anarchism (for one notable exception see Breitbart 1975). Rather, when geographers talk about anti-statism today they usually do so in two ways – as a strategy of neoliberalism, or as a social movement viewpoint.

Geographers define neoliberalism as a form of anti-statism because it is designed to hollow out the state. However, most neoliberals don't want to destroy the state so much as undo its Keynesian form. Neoliberals want to redirect state subsidies from broad public programmes (e.g. food stamps for the poor, road-building for densely populated areas) to private enterprises (e.g. tax breaks for investors). The Keynesian model assumed states could and should manipulate the economy – by adjusting interest rates, establishing a minimum wage, regulating business, etc. It also tried to strike a balance between the competing demands of workers and capitalists. The neoliberal model, by contrast, holds that the state should takes a 'hands off' approach to the economy – limiting regulations, lowering taxes, etc. Moreover, neoliberals believe state spending on social needs is inefficient. The private sector is not only where economic innovation occurs, it is where the greatest returns on investment are found. As such, government investment in the private sector will produce a greater benefit for society as a whole. Ronald Reagan once famously described this as trickle-down economics.

Geographers have examined the anti-statist elements of neoliberalism in both Western and developing economies (see Harvey 2000 and Ould-Mey 1996 respectively). In the Western world they focus on the policies that reorient state functions from public to private interests. In the developing world they examine how international organizations such as the World Bank and the IMF use loan conditionality to force neoliberal 'reforms'.

Geographers also examine anti-statism as a type of social movement (Gallaher and Froehling 2002). Anti-statist social movements can be armed or non-violent. While some use violence, anti-statism is not synonymous with the classic guerilla movement. Guerilla fighters want to overthrow an extant state structure and replace it with another. Anti-statist movements want to limit the influence of any state (Castells 1998). Some want to eliminate the state altogether. Others call for autonomy from it. Still others believe local forms of power should trump state/federal-level power. Most geographers see these movements as responses to neoliberal policies. That is, as the state pulls back on public commitments, citizens feel less connected to the state and more willing to sever formal connections to it (refusing to pay taxes, establishing extra-judicial bodies, etc.). Geographers have also debated whether anti-statism is a supportable position in contemporary society (see especially a special forum in *Political Geography* on anti-statism: Kirby 1997; Luke 1997; Steinberg 1997; Tabor 1997).

Case Studies: Baltimore and Central Kentucky

Neoliberal Baltimore

David Harvey's (2000) study of deindustrialization in Baltimore, Maryland, provides an example of geographic work on the anti-statist policies of neoliberalism. In particular, Harvey examines how the state shifted from meeting public needs to assisting private interests. Harvey describes his adopted hometown as 'a mess'. Between 1960 and 2000, the former shipbuilding hub lost as many as 100,000 manufacturing jobs, with 20,000 disappearing between 1980 and 1985 alone. The city's population declined rapidly as well; a third of its citizens were lost in as many decades. Tuberculosis and HIV infections in Baltimore are now on a par with those of developing nations, as is the city's infant mortality rate.

The city of Baltimore adopted a neoliberal response to the crisis. Instead of focusing on meeting growing public needs (for jobs, food stamps, medicine and affordable housing), the city invested money in private sector development. In particular, the city helped fund a tourist-oriented 'festival marketplace' development in its old shipyards. The 'festival marketplace' concept was developed by urban planners in the 1970s in response to the urban blight associated with deindustrialization. It urged mayors and city councils to use public–private partnerships to rejuvenate downtown and former industrial hubs by turning them into tourist destinations. Planners argued that a festival marketplace would not only refurbish worn-down neighbourhoods, it would provide jobs for former factory workers and increase tax revenue for the city. Festival marketplaces normally include a central mall, cultural destinations such as museums or aquariums, entertainment venues like restaurants and bars, and hotels. Festival marketplaces play off local themes in their design elements. City planners in Baltimore applied the scheme in the city's old shipbuilding corridor, dubbing the development 'Inner Harbor' (Figure 22.2).

Harvey dissects the neoliberal claim that state-assisted private development benefits the public at large. He notes that the city has spent more on Inner Harbor than it has gained in returns. The public–private partnerships that funded the Inner Harbor development are a case in point. Harvey notes they were highly uneven: 'the public takes the risks and the private takes the profits' (2000: 141). The Hyatt Hotel, for example, was lured to the city by a generous contribution from the city of $34.5 million. Indeed, the company spent just $500,000 to construct the Baltimore hotel. Likewise, the Columbus Science Center, a museum built as part of the original festival marketplace design, was constructed using publicly secured loans. When the Center failed, the city was left to foot the bill. The city also gave $2 million in tax relief to a private high rise condominium project to keep it afloat (see Figure 22.3). Harvey likens the city's tourist-based development programmes to 'feeding the downtown monster', noting that 'every new wave of

Figure 22.2 In the mid to late 1980s the city of Baltimore, Maryland, adopted a festival marketplace, turning the city's old shipyards into "Inner Harbor" – a tourist and entertainment centre. David Harvey suggests that Inner Harbor's private businesses were built using an inordinate share of public monies.

public investment is needed to make the last wave pay off' (2000: 141). He concludes by asking why city money should be used to secure private investments that offer little in return to the city. The jobs created by the Inner Harbor, for example, are largely low-end retail work. They pay less than half as much as the jobs they were meant to replace, and they carry limited health care and retirement benefits. To fund the Inner Harbor, the city also cut its services to poor residents, who live in devastating circumstances, surrounded by violence and disease, and with lowered life expectancy.

The people who benefited the most from the Inner Harbor festival marketplace were its private developers, who received generous subsidies, tax relief and in some cases actual bail-outs. In short, while Inner Harbor is scenic and enjoyable, it provides little to the city residents whose tax dollars supported it. The state, in this case a city government, has essentially become an extractive agent of capital, taking public moneys and using them for private purposes. The

Figure 22.3 The high rise, luxury condominium tower Harborview stayed afloat with a $2 million tax relief plan from the city of Baltimore.

state still exists, but its public function has been replaced by a private one.

The Militia Movement in Central Kentucky

The author's work on the militia movement in Kentucky provides a window into an anti-statist social movement in the United States (Gallaher 2003). At the heart of the militia movement is a central paradox: why does a movement that calls for the destruction of the federal government considers itself patriotic?

The rise of the militia movement is related to the application of neoliberal policies in rural America (Dyer 1998; Gallaher 2003). In the early 1970s President Richard Nixon decided to modernize agriculture.

He ordered the Department of Agriculture to secure private loans for farmers to expand their acreage and use of technology. Interest rates were low, and banks, who knew the loans were secured by the government, relaxed standards for borrowers. Farmers signed up for the loans, which carried floating interest rates, in droves (Dyer 1998). The result was an agricultural land bubble, where excess demand drives prices beyond their real value.

In 1979, the land bubble burst when Paul Volcker, then chairman of the Federal Reserve, instituted the first in a series of interest rate hikes to combat inflation. Because farmers' loans had floating interest rates, monthly payments increased by as much as 6 per cent over the next few years. At the same time, the value of their land declined, putting pressure on the banks to collect. By the middle of the decade, thousands of farmers were on the verge of default. The Reagan administration's response to the crisis in rural America was to take a hands-off approach, arguing that the market was adjusting itself and that government intervention would only create inefficiency in the agricultural sector. By the end of the 1980s almost one million family farms were foreclosed (Dyer 1998).

In response to the farm crisis white supremacists such as David Duke, a former grand dragon of the Ku Klux Klan, and well established racist groups like the *posse comitatus* began organizing distressed farmers (Levitas 2002). They repackaged their racist message (against Jews and other minorities) in anti-government rhetoric. They told farmers that international bankers were working with the federal government to steal their land.[1] They also claimed that the US government had been overtaken by communists who wanted to create a one world government that would transfer American wealth to undeserving people in the Third World.

Activists gave farmers 'solutions' – they taught them how to put liens on the property of people involved in the foreclosure process (bankers, judges, lawyers), how to create and run 'common law' courts, and how to form militias to save their farms on foreclosure day. These solutions were illogical and in many ways counter-productive (several farmers cum militia men died in shootouts with law enforcement agents while resisting foreclosure). However, they provided desperate farmers with an identifiable enemy – the state – and a deceptively simple solution: to pretend agents of the state (foreclosure judges, local sheriffs, etc.) were illegitimate and could be ignored.

After the farm crisis subsided in the late 1980s, activists turned their efforts to areas not directly affected by the crisis, such as the

Figure 22.4 A flyer for sale at a militia meeting in Garrard County, Kentucky, 1997. The militia movement argues that government agencies such as the Bureau of Alcohol, Tobacco, and Firearms (BATF) abuse their power against American citizens.

inter-mountain west and the upper South. In Kentucky, a militia formed in the early 1990s. The group, which eventually became the Kentucky State Militia (KSM), concentrated its efforts on gun laws. They pointed to government excesses at Ruby Ridge and Waco as evidence that the government could and would use force against its own citizens.[2] An armed citizenry, they argued, was the last line of defence against federal tyranny (see Figure 22.4). The group's first achievement was to help get a 1995 law passed to allow the state's citizenry to carry concealed weapons.

As the group consolidated, it expanded its activism to other issues. For example, the Kentucky State Militia began protesting against the designation of the Land Between the Lakes region of the state as a biosphere reserve. The biosphere designation is given to protected bio-geographical regions across the globe by the United Nations Educational, Scientific and Cultural Organization (UNESCO). The land, which is federally owned and managed, drew the ire of KSM, which saw sinister motives behind the UN designation. The group issued press releases to local news outlets suggesting the UN owned the land and was practising troop manoeuvres in a bid to take over the southern United States. It also alleged that the federal government was in on the plot, essentially allowing the UN to prepare for an imminent take-over. Although neither claim was true, KSM insisted the US government was working in concert with the UN to rob ordinary Americans of their land and give it to 'undeserving' people from the Third World (Gallaher 2003).

KSM's anti-statism was the product of deep anxiety born of neoliberalism. Deindustrialization and agricultural restructuring have left white rural, working-class men economically and culturally adrift. These men find it increasingly difficult to secure breadwinner jobs, and their dominance in the social hierarchy is put in question as a result. With the state planted firmly on the side of capital, the militia movement has rejected the state. The role of racist activists in the movement's beginning has also allowed these men to project all sorts of conspiracies onto the state.

KEY POINTS

- Anti-statism is a term used to describe opposition to the state and its power to regulate social, economic and political life.
- Anti-statism can emerge from above and below.
- Anarchism is one form of anti-statism. Anarchists believe that humans are inherently cooperative and function best under decentralized forms of governance. Pyotr Kropotkin and Élisée Reclus are geography's most famous anarchists.
- When geographers talk about anti-statism today they usually do so in two ways: as a neoliberal strategy to reduce state influence over the economy, or as social movement protest against federal power.

NOTES

1 In an effort to increase recruitment, activists toned down their rhetoric. In earlier iterations, for example, 'international bankers' would have been described as 'Jewish bankers'.

2 Ruby Ridge and Waco refer to two standoffs between armed federal agents and American citizens in the early 1990s. In both standoffs armed agents killed innocent civilians. At Ruby Ridge, Idaho, US marshals surrounded the property of Randy Weaver, who had failed to show up for a court hearing related to his indictment on federal gun charges. The siege ended with the death of one federal marshal, Randy Weaver's wife, and his 14-year-old son. A year later the government laid siege to the Branch Davidian compound outside of Waco, Texas. The group's leader, David Koresh, was wanted for stockpiling weapons. The government used a 'dynamic entry' method to issue Koresh's arrest warrant. The Davidians were tipped off about the plans, however, and met the assault with force. Four federal agents and six Davidians were killed. The government then surrounded the compound for 50 days (an immediate response was ruled out because children lived in the compound). When the Attorney-General gave the go-ahead to forcibly end the siege, the compound caught fire, killing 74 occupants, many of them children. The fire is believed to have started when the government fired CS gas into the compound to force people out.

FURTHER READING

Gallaher, C. (2003) *On the Fault Line: Race, Class, and the American Patriot Movement*. Lanham, MD: Rowman and Littlefield.

Kropotkin, P. (1885) 'What geography should be', *Nineteenth Century*, 18: 940–956.

Ould-Mey, M. (1996) *Global Restructuring and Peripheral States: The Carrot and the Stick in Mauritania*. Lanham, MD: Littlefield Adams Books.

Reclus, É. (1991) *L'Homme et la terre: Histoire contemporaine*, Paris: Fayard.

Part VI
Identity

INTRODUCTION

Carolyn Gallaher

In the last twenty-five years, issues of identity have become a central concern in the discipline of geography. The focus on identity, generally, and identity politics more specifically, is related in part to the changing political landscape. The 1960s saw the birth of a variety of identity-based social movements, including feminism, black power, gay rights, political Islam, the religious right and ethnoregionalism (e.g. Scottish nationalism). These movements could not be easily described or explained using standard categories of analysis such as class. Over time, social scientists in (and out of) geography developed new theories and concepts to explain the rise of identity-based forms of politics. For their part, political geographers tend to examine the spatial qualities of identity politics (that is, how identity is spatialized and how space is used to rally support for political rights for specific identity groups).

When political geographers study identity politics, some of them rely on standard concepts, but update them to account for identity politics. For example, **nationalism**, the subject of Chapter 23, can now refer not only to identity politics of extant nation-states, but to sub- and supranational groups. Likewise, the concept of **citizenship**, the subject of the second chapter in this part, is used to refer not only to one's attachment to a nation-state, but also to places within nations, or those that cross the boundaries of several nations. Moreover, in an increasingly globalized world, it is possible to find people who have nationalistic sentiments about more than one place and who can claim citizenship in multiple states as well.

The third chapter in Part VI examines identity and identity politics with reference to the postcolonial world. The study of **postcolonialism** has import not only in formerly colonized places, but also in parts of the world where significant numbers of people from formerly colonized states have migrated. The migration of people from the former British West Indies has left an indelible mark on British culture. In political geography postcolonialism is associated with critical geopolitics and feminist approaches. As such when political geographers study the postcolonial condition they often note the continuation of colonial-like practices after the end of formal colonization. They also postulate ways in which such practices can be resisted.

Chapter 26 deals with the issue of **representation**. As I note in the Introduction to this volume, the concept of representation has traditionally covered the mechanisms by which a citizenry is represented in local, provincial and national governments. In this book we cover more recent understandings of representation, which deal with the ways that identity-based groups express their identity, culture and political views. Political geographers examine how such groups represent themselves; how dominant groups use representational tropes to keep such groups 'in their place'; and how these groups resist such representations.

Chapter 27 covers the concept of **gender**. The emergence of feminism in the 1960s taught us that one's gender, and problems stemming from the normative restrictions attached to it, can lead to social activism and social change. Political geographers note that gender is a social construct, as are the norms built around gender roles. That is, while a person may have male sexual organs, the roles for 'proper' behaviour for males vary across time and space. In political geography feminists argue that many of the subdiscipline's key concepts are gendered. They note, for example that political geography has tended to ignore the political relations that occur at smaller scales where women tend to have more say.

The final chapter details the concept of **'the other'.** In many ways the idea of the other forms the basis of most theories of identity. In brief, identity theory hold that any normative identity (i.e. the dominant identity in a society) requires an other for its formation. That is, one cannot be white unless there are other people who are not white. Social scientists interested in identity theory have tended to examine how 'othered' groups are oppressed and how they resist that oppression. More recent analyses have also focused on the ways that dominant identities are normalized such that their central place in society is unquestioned, accepted as 'normal' and right.

The case studies provided in this part of the book are as diverse as those in other sections. Social movements such as Las Madres in Argentina are discussed. So too are ways in which British Muslims interact with the country's dominant Christian culture. Street level conflict in Belfast is also detailed with an eye to the ways in which these conflicts are represented in paramilitary murals.

23 NATIONALISM

Alison Mountz

Definition: An Elusive yet Powerful Force

Nationalism usually refers to the territorial expression of identity: a sense of belonging to a group or community that shares a common identity, often but not always associated with a particular territory. Defining a nation is no simple matter. A shared national identity is often defined by belonging to a territory or country with a common political system. In other cases, however, national identity may reference belonging to a group with a shared set of beliefs, faiths, traditions or even linguistic practices that transcend national borders and political systems. Nationalism is an elusive yet powerful concept: difficult to define and locate, but impossible to ignore.

The birth of the modern nation-state is most commonly associated with the Westphalian system originating in 1648. Westphalian sovereignty assigned the power of domestic governance to individual states, free from intervention by other nation-states. This independent governance would therefore correspond with the territorial limits of the state. Whereas nation-states are a relatively recent entity in world history arriving as they did in the late seventeenth century, they have proven a very powerful force in contemporary life. The location where we are born, and the people to whom we are born, influences every aspect of contemporary life, from what we do, where we travel and our participation in economic and political systems, to formal citizenship, an informal sense of belonging, loyalties and views of the world. These modes of belonging to a nation-state potentially influence a person's service to and indeed subversion of their country (for more information on the concept of the nation-state, see Chapter 1). Nationalism is one mode of belonging.

Benedict Anderson (1983) identified a subsequent phenomenon, tied to yet independent from, the birth of the nation-state: the development of nationalism during the nineteenth century. He explored notions of identity and belonging that arose in and corresponded with the territory of the nation-state. Anderson famously declared that nationalism was the operation of an 'imagined political community' in which people cannot possibly all know one another, but believe themselves to hold certain histories, narratives and dreams in common. Such groups also have shared symbols, iconography, memories and structures that act as unifying forces. They share these beliefs through shared simultaneous events and their telling, as in the daily act of reading the newspaper.

Nationalistic sentiments will influence how a collective group functions. Multiple forms of nationalism often exist and even compete within the same national territories. Competing histories may contradict one another and cause conflict. Nationalism thus operates as a powerful force not only in the everyday lives of individuals, but in those of nation-states as well. Eric Hobsbawm (1990) argues that governments often hold different perspectives on nationalism than do members of civil society. Such competing ideas of nationalism can lead to civil unrest and even civil war. Political geographers tend to think about nationalism as a force operating on and within a nation-state. They call these centripetal and centrifugal forces. It is thought that too many nationalisms operating within one state may destabilize that state, or pull it apart by operating as centrifugal forces. Conversely, centripetal forces work to pull a state and its corresponding national community together. In the 2008 campaign to be the Democratic nominee for presidential election in the United States, Senator Barack Obama attempted to capitalize on centripetal forces. His campaign platform called for a time of national unity to overcome forces such as race and partisan politics that had divided US citizens.

Political geographers often study cases where national identities do not correspond with the territory of the nation-state. Indonesia, for example, is a nation-state comprised of many islands, peoples, religious and linguistic communities. These differences have historically acted as centrifugal forces that threaten to pull apart the state. The Indonesian state frequently has to quell civil unrest and independence movements, for example the West Papuan independence movement. Another case where a nationalist group has called for independence can be found in the Canadian province of Quebec, whose inhabitants have periodically sought sovereignty, sometimes through violent means. The most recent attempt, a referendum vote in 1995, failed narrowly (Figure 23.1). This nationalist group's collective identity did not correspond with that of the nation-state of residence.

Figure 23.1 The Quebec flag painted on a young boy's face. In 1995 the citizens of Quebec voted on a sovereignty referendum. The measure narrowly failed, leaving Quebec as a province within Canada.

Evolution and Debate: Complex Boundaries

Nationalisms grew in the eighteenth and nineteenth centuries, organized around popular ideas of membership in nation-states. Many people were wary of the rise in nationalism and wanted to see alliances based on other forms of identity, such as class or even geographical regions larger than nation-states. Modernists believed that place-based identities would ultimately give way to global identities; that local affinities would fade with globalization and the ability to travel and communicate quickly across international borders (see Harvey 1989 for further discussion).

But the opposite turned out to be true. Since 1945, nationalism has thrived as nation-states have proliferated. In 1940, there were 65 nation-states. There are now over 190. This growth resulted in part from decolonization, particularly in regions of the global south where the nation-states mapped by colonial powers did not correspond with ethnocultural, tribal, pan-religious, linguistic or faith-based communities and *their* boundaries. Further proliferation of states occurred with the dissolution of the Soviet

Union in 1991. A corresponding phenomenon has been the growth in stateless persons, when people found themselves living in a newly formed state where they did not feel a sense of belonging or were not accepted as belonging.

A more contemporary argument suggests that nationalism arose in relation to globalization and global notions of culture (e.g. Featherstone 1990; for more details, see Chapter 14 on globalization). At the same time that globalization brings us all closer together in an intimately chaotic global world, we seek shared ethnic and national identities in order to narrate who we are and how we are different from one another.

Some national communities have successfully established states that express their common identities, as in the establishment of Israel as a nation-state and homeland for the Jewish diaspora in 1948. More commonly, however, nationalisms have corresponded less over time with the boundaries of nation-states as groups identify as nations without sovereign state territory. Those which have invested in a pan-Arab form of nationalism, for example, identify regionally with a territory that transcends the borders of modern nation-states. In short, groups that identify in nationalistic terms do not necessarily attach this belonging to a geographical territory demarcated by the borders of a nation-state.

To further complicate matters, many multinational states exist. These states govern multiple nationalities in one political unit where languages and ethnicities proliferate. Examples include Switzerland, the United States, South Africa and Canada. Among ethnic or cultural groups that inhabit multinational countries, some will seek or have recognition by the state, while others might embark on movements to secede or remove themselves from the state to form their own territory. Canada is an example of a country with multiple nations and nationalism where peace prevails. Although 18 per cent of the population was born outside of the country, multiple national identities, ethnicities and languages co-exist peacefully and are recognized by the federal government, which prides itself on a federal policy of multiculturalism.

Yet ethnic nationalism continues to rise globally. In some multinational states, nationalist groups are seeking independence as separate states, sometimes by violent means. In Sri Lanka, for example, the Tamil population has experienced decades of discrimination by the reigning Sinhala majority. Since 1983, the Liberation Tigers of Tamil Elam (LTTE) have been fighting for a separate Tamil homeland as a solution to this history (Figure 23.2).

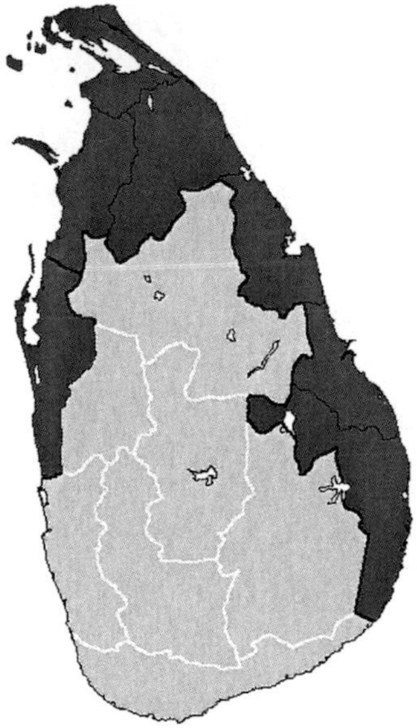

Figure 23.2 Map of Sri Lanka: shaded area is claimed by the Tamil Tigers (as of August 2005). Some of the areas claimed by the Tigers are not under their control.

Nationalist movements arise in the searches not only for recognition of identity but for fuller participation in the economy in response to uneven economic development and inclusion. Globalization does not have even effects across nation-states or regions. Often regions lagging in economic development are those with a concentration of ethnic minorities. Ethnic groups might also be over-represented in segments of the labour market, providing fuel for political mobilization in the belief that members will do better as an autonomous entity than they will as a colonized or subjugated people. Leaders and other agents of change take hold and propel a movement forward, drawing on those shared histories, languages, beliefs and symbols that pull them together, in the name of preserving the past and the future. States, too, often co-opt and incorporate narratives of nationalism into stories of modernity (e.g. Nelson 1999).

Some states will attempt to accommodate, incorporate, even manage multiple nationalisms, as in the Canadian case. Others will not succeed, as in the case of the Tamil Tigers and that of the Chechen nationalists detailed below. Thousands of deaths around the globe have resulted from the power of nationalism to mobilize political action. Nationalist movements led to the end of the Soviet Union, Yugoslavia and Czechoslovakia; all splintered into multiple nation-states.

This very brief historical sketch of nationalism suggests a key role for the study of geographies of nationalism. And indeed, geographers have been central to conceptual understandings of nationalism and in particular as to how these relate to state-building projects and social movements.

Case Study: Nationalism, Terrorism and Government Responses

Nationalism among a population excluded from economic development can evolve into terrorist movements that, in turn, prompt powerful enforcement responses by nation-states.

Nationalism references not only the territory of the nation-state, but spatiality more broadly, temporality, body politics and notions of family (usually conceptualized in patriarchal terms) (Kaplan et al. 1999: 15). While by no means the exclusive purveyors of national identity, national governments and leaders often attempt to harness the power of nationalism in order to consolidate power in response to terrorist acts. Following the terrorist attacks by Al Qaeda in New York and Washington, DC, in September 2001, the US was awash with patriotic fervour. Young men signed on for military service, and flags proliferated, posted everywhere from the bumpers of cars to private residences and places of business. President Bush capitalized on this nationalism and harnessed its discursive power in the march to war, first in Afghanistan ostensibly to capture Al Qaeda's leaders, and then in Iraq.

At home, President Bush centralized various federal agencies in the newly created Department of Homeland Security, the largest change in the bureaucratic structure of the US federal government since President Franklin D. Roosevelt's New Deal of the 1930s. New legislation on terrorism was passed in several countries, and in the United States this took the form of the Patriot Act. The very name of the Act references nationalist projects in the name of fighting terrorism, and indeed the Act, signed into law in 2001, granted more extensive powers to authorities to conduct surveillance on foreign and national citizens at home and abroad.

Before its dissolution the USSR had hundreds of nationality groups, all recognized as separate nations by a central government. The central government conducted censuses to determine national identity, which was included on internal passports. In 1991, the USSR was devolved into 15 republics. Nationalities then corresponded with spatial units, but conflicts related to nationality and citizenship remained. Groups were scrambling to move or to assimilate into new nation-states by learning new languages. People had to resolve questions of nationality and citizenship and in the process statelessness became a significant problem.

Some countries have attempted more federalist structures, adopting a power-sharing model characterized by more diffuse decision-making, with the aim of incorporating autonomous groups into the governance of the nation-state. Russia pursued this goal, granting relative autonomy to its 31 ethnic republics. After the dissolution of the Soviet Union and the ensuing conflict over ethnicity and independence, Chechnya (or Ichkeria to inhabitants) fought for and won *de facto* independence – but not *de jure* independence. Civil war over independence broke out in 1992 and people of non-Chechen ethnicity fled. Russian forces entered Chechnya in 1994, beginning the First Chechen War (Figure 23.3). In 1996, after over 100,000 deaths, Russian forces withdrew (Figure 23.4).

Chechnya did not become its own nation-state but remained a part of Russia. Some speculate that this relates to the location of oil fields near Chechnya and the desire for Russia to maintain access to these oil fields, and therefore political and economic power in the region.

But this has not been a peaceful arrangement. Cleavage usually results from tension that pits a national core where political power, culture, urban elites and wealth amass (as they have quite dramatically in Moscow in recent years) against more peripheral districts. Chechnya is clearly on the periphery in Russia, not only geographically, but in terms of its marginalization from political, cultural and economic power.

Many Chechens felt themselves to be part of their own nation. Nationalisms, however, were suppressed under the imperial rule of the Soviet Union with the denial of territorial autonomy in the form of statehood. Nationalism was used as a tool to control rather than empower residents of the Soviet Union. But with the breaking apart of communist rule, many of those identities that had been suppressed are now emerging, some more powerfully and others more fraught. Chechen rebel groups embarked on terrorist acts in order for their discontent to be heard. In 1999, Russian forces re-entered Chechnya, beginning the Second Chechen War.

Figure 23.3 Russian helicopter downed during the First Chechen War (Photo by and courtesy of Mikhail Evstafiev)

Increasingly, some subgroupings of nationalists have turned to terrorism in an attempt to cause enough commotion to realize the dream of a nation-state. These are cross-border acts of violence with some transnational ties, more suspected than proven, although at very least following some of the tactics of Al Qaeda. The groups follow clear strategies of terrorism to disrupt daily activities. These have included the bombing of people in their apartment buildings, the hostage-taking of those attending an evening theatre event in Moscow, the bombing of a morning metro commute to work and a successful attack on two airplanes in flight.

In 2004, Chechen rebel groups in Beslan, a small town, laid siege to Middle School Number 1 as children and parents attended opening ceremonies on the first day of term. Over 1,100 children and their parents were held hostage in the school gymnasium for three days while negotiations with authorities ensued. The standoff ended with the death of 334 civilians, more than half of them children.

Much information regarding the hostage-taking at the Beslan School and its conclusion remains disputed, in spite of various commissions to

Figure 23.4 Chechen Fighter during the Battle for Grozny (Photo by and courtesy of Mikhail Evstafiev)

look into what happened and to assign responsibility. In subsequent years, the Mothers of Beslan have continued to demand the release of information and video footage from authorities.

Contemporary feminist scholarship examines the intersection of gender and nationalism, showing how men and women have been affected in different yet related ways by conflict (e.g. Giles and Hyndman 2004; Kaplan et al. 1999; Mayer 2004). This scholarship links violence to the female body and draws connections between the roles played by men and women in family and nation (Gilbert and Cowan 2007; Nelson 1999).

Russia was led by steadfast President Vladimir Putin, who had refused to negotiate with terrorists or separatists. In response to this and earlier attacks, Putin centralized power in an overhaul of the political system in order to consolidate power centrally and aggressively suppress political dissent (Myers 2004). He too drew on the discourse of nationalism by declaring the need for a time of national unity rather than ethnic division in order to fight terrorism. In a speech shortly after the siege, Putin remarked that 'Those who inspire, organize and carry out terrorist acts are striving to disintegrate the country' (Myers 2004). Like Bush, Putin became more isolationist in response to terrorism, consolidating his own power in the executive branch over legislative and regional branches of government. He expanded the powers of law enforcement agencies and enhanced security measures in cities.

Bush and Putin pursued strategies that sought the consolidation of power in the name of national unity in the fight against terrorism. But pro-democracy activists in both Russia and the United States were quick to criticize their decisions. Those in the United States argued that ethnic and racial profiling were no way to advance border enforcement and immigration policies. And in Russia it was argued that efforts to erase ethnic division in the name of an ethnically united nation had already failed.

Future Research on Nationalism

Political geographers must continue to address important questions about nationalism. How do nation-states balance the rights of minority groups with a sense of national identity and stability? The stakes are high, as failure may result in civil unrest, or even genocide in more extreme cases. In the parlance of political geographers, is nationalism a *centripetal* force that holds the state together, or a *centrifugal* force that pulls it apart? What circumstances cause nationalism to shift from a centripetal to centrifugal force? At what point can desires for separate territory and identity evolve into terrorist movements? And how do states respond to these efforts? As globalization continues in the form of cross-border flows, will we continue to identify ourselves with groups that are territorially based?

National movements raise a number of challenges, particularly for multinational nation-states facing separatist movements. What happens when nationalist movements inspire terrorist acts that threaten national security? Whose nationalism will be respected, and whose security protected? Do measures taken in the name of national security put particular groups of people at risk, trading human security for national security? Answers to these questions will vary among persons residing in Israel or Palestine, Northern Ireland or England, Chechnya or Moscow. Because location – geography – matters.

KEY POINTS

- Nationalism refers to the territorial expression of identity: a sense of belonging to a group or community associated with a particular territory.
- Political geographers often study cases where national identities do not correspond with the territory of the nation-state. Many people in the Canadian province of Quebec, for example, view Quebec as a distinct nation and believe it should become its own nation-state.
- Most states in the world are multinational states, meaning that they contain a variety of nationalities.
- Nationalist movements are on the rise globally. Many of these movements use violence, or the threat of violence, to secure independence, separation or autonomy from an extant nation-state.

FURTHER READING

Anderson, B. (1983) *Imagined Communities*: Reflections on the Origins and Spread of Nationalism. London and New York: Verso.

Hobsbawm, E. (1990) *Nations and Nationalism since 1780*. Cambridge: Cambridge University Press.

Sparke, M. (2005) *In the Space of Theory*. Minneapolis: University of Minnesota Press.

24 CITIZENSHIP

Alison Mountz

Definition: Spatial Belonging

Citizenship is one among many forms of belonging. Most commonly, it means belonging to the nation-state. The term is widely believed to have its origin in ancient Greek forms of participation in the city. While the history of citizenship as a practice is subject to debate (e.g. Isin 2002), invariably the term has meant membership in relation to territory. (For more information on territory, see Chapter 6.) Territory here includes a wide array of sites, scales and spaces from the local city or city-state to the modern nation-state. Contemporary notions of citizenship are also cast globally, as the term 'global citizen' suggests.

Rights and claims to citizenship are most often regulated by nation-states and based either on place of birth (*jus solis,* Latin for right of the soil or birthright to territorial belonging) or parentage (*jus sanguini,* Latin for right of blood, wherein citizenship is handed down through children). Any person born on US soil, for example, is a US citizen on *jus solis* grounds. The children of a Canadian parent born abroad will have rights to Canadian citizenship by virtue of *jus sanguini.* Claims to citizenship can also be made on other grounds. Jews, for example, regardless of place of birth, have a right to Israeli citizenship by virtue of their religion.

Some persons immigrate to a country to work and live; others arrive as refugees in search of protection from prosecution at home; still others may make claims based on lineage. Most will apply for citizenship and undergo a process of naturalization that varies from state to state.

The nation-state has long held the power to determine and regulate citizenship as legal membership or belonging. Recently, however, scholars have grown more interested in contemporary practices of citizenship. Such practices are often deterritorialized, with scholars moving their focus away from legal attachment to a homeland and towards daily participation in society. Aihwa Ong (2006: 7) suggests that 'components of citizenship have developed separate links to new spaces, becoming rearticulated, redefined, and reimagined in relation to diverse locations and ethical situations'. In a global era, people seek and find modes of belonging that extend beyond the geographical delineation and legal regulations of the nation-state.

In the formal and legal sense, citizenship influences human behaviours in relation to rights afforded by and responsibilities to the state, such as the right (and sometimes the responsibility) to vote. Citizenship structures a person's mobility and serves as a basis for protection against threats to life or livelihood. The rights of most citizens are more ardently protected by nation-states than are the human rights of 'non-citizens' (Arendt 1958).

In the more informal sense, citizenship references a range of political practices, and a vast array of modes of participation in civil society have emerged as part of this concept. Some study forms of protest and activism, for example, as practices of citizenship. These have become important topics of study among political geographers (e.g. Brown 1997; Mitchell, D. 2003; Routledge 2003b).

Evolution and Debate: A Concept Redefined by Globalization

With the development of the modern nation-state, citizenship became a formalized way for states to take a census of population and ultimately categorize and monitor membership. This formal kind of belonging is attached to a particular place: the territory of the nation-state. Citizenship means membership in a society; this membership is documented by the nation-state for a variety of purposes, such as the regulation of mobility (Hannah 2000). By virtue of their documentation, citizens are ordered and rendered legible by the state (Torpey 2000).

John Torpey (2000: 4) suggests that 'people have become dependent on states for possession of an "identity" and offers a typology that involves three kinds of documentation associated with nationality. First, international passports enable people not only to come and go from their own nation-state, but also to enter the territories of other nation-states as well. Second, 'internal passports' regulate human migration *within* one nation-state, a form of regulation largely abandoned today as it is considered an antiquated practice associated with authoritarian regimes (Torpey 2000: 165). The pass laws instated in

Figure 24.1 California Driver's licence. In most American states only citizens and permanent residents are permitted to apply for and obtain a driver's licence. Not having a driver's licence in the US can mark you as 'not belonging.'

South Africa in 1923 under apartheid, for example, restricted and regulated the mobility of black South Africans in cities. Third in Torpey's general typology are identity documents. In many countries, a citizen can be called upon by government authorities (police, court clerks, etc.) to provide documents, such as a driver's licence or other government issued document, to verify his or her identity and citizenship (Figure 24.1).

Citizenship, in practice, has become more 'flexible', as Aihwa Ong (1999) argues. People that Ong identifies as 'flexible' now have multiple passports in order to move frequently between countries to work. Some are even members of 'astronaut families' that live in multiple countries (Waters 2002). Ong demonstrates, however, that such degrees of international mobility are regulated in relation to a person's class, ethnicity and location. Wealthier individuals and families have more money to travel and access to the political channels that enable them to obtain visas and jobs abroad. These global citizens are more likely to be granted what Ong terms 'flexible citizenship'.

Globalization and the increased pace of human mobility across borders that characterized the late twentieth century have challenged traditional notions of citizenship. People are on the move internationally, often many times a year. They participate in households and jobs that span national borders and territories. In order to capitalize on these transnational practices of citizenship, some states have formalized the ability of individuals to hold multiple forms of citizenship.

Contemporary scholars of citizenship examine not only the philosophical underpinnings and historical referents of formal citizenship, but also differential practices of citizenship along other axes of difference, such as gender, age, ability, sexual identity, race, ethnicity, country of origin and occupation (see Appadurai 1996; Ong 2006, 1999; Pratt 2004). Important examples throughout history demonstrate the ways that populations were discriminated against and subject to control on the basis of identities determined and enforced by state authorities. Engin Isin (2002) argues that citizenship as a mode of formal belonging is indeed defined by those constituted outside, beyond the limits of citizenship. In South Africa, for example, non-white citizens were literally segregated and contained in townships, their mobility and citizenship strictly limited by the smaller, wealthier white population in power under the system of apartheid. Today, groups of migrant labourers from the global south are often discriminated against in the countries of the global north where they are welcomed as low-wage workers but not as legal citizens with the formal status and rights accorded to citizens. Citizenship practices in North America and the European Union privilege those with particular sets of skills and educational degrees, such as doctors or IT workers, over those who come to work in agriculture, construction or the low-end service economy. Similarly, formal citizenship practices in immigrant and refugee-receiving countries privilege certain countries of origin over others. In the United States, for example, it has historically been easier for persons fleeing communist regimes such as Cuba to receive refugee status than for those persons fleeing non-communist regimes supported by the US, such as El Salvador. Refugee policy and access to citizenship thus mirror US foreign policy and geopolitical relationships.

Many states now accept 'dual citizenship' wherein a person can 'claim allegiance', belong to and hold passports in more than one country. Some transnational migrants even hold three or four passports. Government authorities, meanwhile, worry over and debate the extent to which multiple sites of belonging will influence a person's social and economic contributions to a place.

These formal, legal changes in the regulation of citizenship have been accompanied by informal changes in cultural practices and people's sense of belonging. A person may, for example, hold citizenship in one country but feel more at home in another. Individuals and communities of people have developed their own cultural forms of citizenship alongside the more legalized forms of citizenship governed by nation-states. Isin (2002) offers an open interpretation of the range of behaviours referenced by the term 'citizenship', arguing that citizenship is always defined in order to narrate normative and non-normative or deviant behaviour.

Still others are marked and excluded, even violently persecuted, on the basis of other aspects of their identity. During the Holocaust, Jews and sexual minorities were forced to wear patches on their shirts for the purposes of identification, imprisonment in work camps and death in concentration camps (see Figure 24.2). Today, discrimination occurs, sometimes in a more subtle if insidious fashion, around the globe, wherein power works against persons with sex, gender identity, age or ability that operates beyond 'normative' measures and confines.

Case Study: Filipina Immigrants in Canada and Statelessness in Estonia

Filipina Immigrants in Canada

Political geographers have studied citizenship in a variety of ways over time, and like the practices of citizenship they study, have grown more flexible and adopted more cultural understandings of spatial practices and processes of belonging (Ley and Kobayashi 2005; Silvey and Lawson 1999; Silvey 2004). Geraldine Pratt's (2004) work provides a helpful case in point. Her study of the migration, work and living experiences of immigrants from the Philippines to British Columbia, Canada, traverses multiple scales and sites where citizenship is practised, both formally and informally.

Pratt (e.g. 2004; Pratt et al. 1998) spent many years working collabouratively with a community organization in downtown Vancouver, British Columbia, called the Philippine Women Centre. This organization is a grassroots advocacy group for Filipino immigrants in Canada.

Figure 24.2 Concentration camp overalls with a Star of David patch.

Some involved are Filipina-Canadians who were born in Canada, while others immigrated to Canada from the Philippines. Pratt joined the organization in devising strategies to advocate for better working conditions and better paths to citizenship for live-in caregivers, and she has written collabouratively of this work (Pratt et al. 1998).

One subset of the immigrants detailed in Pratt's study came to Canada through the Live-in Caregiver Programmes, a special and temporary two-year immigration programme devised by the government to recruit home caregivers for children. The programme facilitates the migration of Filipino workers, most of them women, to live with and work for a family. If the worker maintains steady employment for two years, then she is eligible to apply for 'permanent residence' in Canada, the next step on the road to Canadian citizenship. At that point, she acquires more freedom and mobility and the right to work in any job.

Pratt argues that the requirement that caregivers remain with the family for two years without seeking an education or working elsewhere has structured into the citizenship process the conditions for abuse of workers. These women often lack time off from their jobs and privacy within the homes where they live and work. Because they are required to live in the home of their employer, and because child care requires continuous work throughout the day and night, live-in caregivers are often exploited. There is no clear beginning or end to their work-day; they never actually 'leave' the space of work.

The Philippine Women Centre has been involved in drawing attention to the plight of women who come to Canada through this programme. Their experiences demonstrate that the dimensions of belonging are formulated as much in the formal legal policies of citizenship – the route to membership in the Canadian nation-state through temporary visa, permanent residency and then naturalization – as they are by informal daily exchanges in the home, the city and the workplace where Filipina domestic workers are often treated as 'second class citizens', quite literally.

Their country of origin influences the work that Filipina immigrants do in Canada, more so than their educational degrees and work experiences in the Philippines. Racialized forms of inclusion through this particular labour market also structure the ways that employers and others interact with the women in private and in public spaces. Though women who immigrate through the temporary programmes may have professional degrees from home, they are treated poorly as temporary labour in Canada. And though many Filipina immigrants and their families enter other professions in Canada, they are stereotypically and homogeneously treated as live-in caregivers.

The two-year programme is an employment contract that capitalizes on cheaper immigrant labour from the global south. As such, the international division of labour facilitates economic lifestyle opportunities for wealthy families of the global north by dividing families of the global south. Differences along axes of gender, age, ethnicity, class, country of origin and legal status support this system. Structural economic disparities between regions prompt young women to emigrate in search of economic livelihoods and opportunities. Their age, ethnicity, gender and legal status in Canada factor into discrimination by employers who abuse them. 'Canada as a society exploits third-world women as a low-cost labour force' (Pratt 2004: 39).

Pratt, 'facing the embodied pain of forced migration' (2004: 5), became interested in the ways that formal modes of belonging were contested and subverted through protest, activism and alternative ways of engaging the Canadian federal government. Often those groups marginalized by the state have led the way in organizing for social change, whether communities of immigrants who call attention to social and economic injustice, such as the Filipina domestic workers, or persons who may hold citizenship status but find themselves discriminated against in practice as a group, as in the case of persons with HIV/AIDS (Brown 1997). In relation to the Filipino community, Pratt (2004) found that many young people who were born in Canada, and in many cases have never been to the Phillippines, have become community activists and leaders in fighting the exploitation of Filipina workers. Their politicization arose from their Witnessing the exploitation of their mothers, aunts and sisters in Canada. This social movement for workers' rights and political inclusion reflects the power of both formal and informal practices and claims to citizenship

Statelessness in Estonia

At the same time that the practice of dual citizenship grows and the bureaucratization of formal citizenship expands unabated, so too does the phenomenon called statelessness, when an individual lacks formal documentation establishing their citizenship of any state in the world. The number of stateless persons increased in the late twentieth century for a variety of reasons. Most famous were those displaced in Europe in World War II, among them Jews born under Hitler's Third Reich but denied citizenship. Others were members of nations in Africa that never corresponded to the national borders drawn by colonial powers. As people migrate, they lose their right to citizenship, since their location does not correspond to authorities' interpretations of belonging. Still others live in states not recognized by other states in the international community, as in the case of Palestinians in Palestine; or they have left states that are not recognized, for example Palestinian refugees in Lebanon who do not have citizenship in Lebanon, but are also not permitted to return to Palestine.

More recently, the breaking apart of the Soviet Union instigated a spate of stateless persons in the 1990s. Some returned to their ethnic homeland and became naturalized there, while others acquired citizenship of the new nation-state in which they were located by residence. This caused much confusion as many residents' self-identified ethnicity and nationality did not correspond with their place of residence.

Estonia provides one such example. In the late 1990s, nearly one-third of its population was stateless, and the government invested heavily in

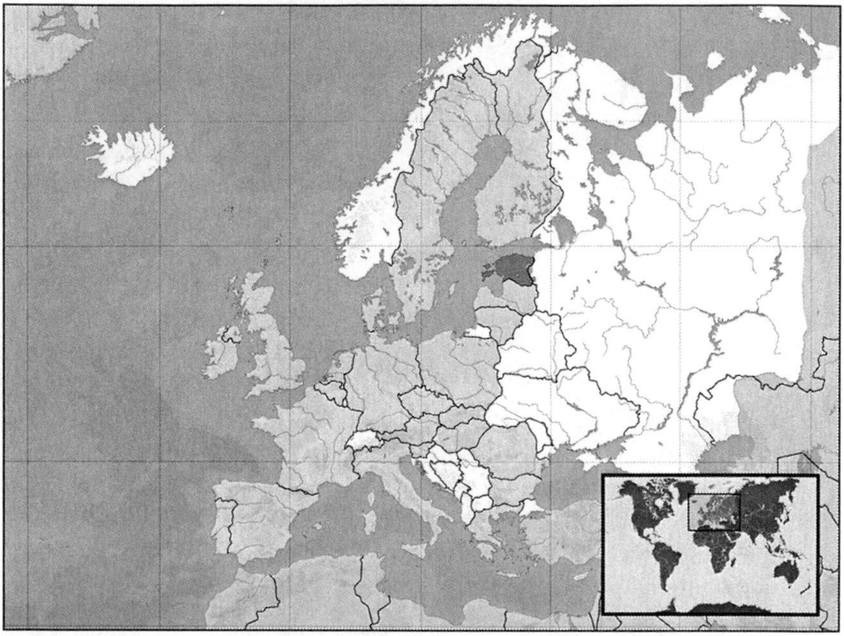

Figure 24.3 Estonia within EU countries as of 2007

a naturalization drive to encourage residents to become citizens of Estonia. Higher rates of citizenship were a requirement for Estonia to join the European Union in 2004, as citizens would then hold EU passports and have more rights to move and work throughout the EU (Figure 24.3). Some who naturalized were Estonians, but many were ethnic Russians living in what became Estonia. Their own cultural sense of belonging did not correspond with the authorities' efforts to govern citizenship as membership in a nation-state (see Feldman 2005).

The novel *Border State* (2000) by the celebrated Estonian author Tonu Onnepalu beautifully evokes the meaning of citizenship for a protagonist who finds himself on the margins of formal and informal modes of belonging. The novel traces an unnamed character who leaves home, an 'underdeveloped' and feminized country of 'the East', to move to Paris, a hypermodern, cosmopolitan city of 'the West'. There, he holds temporary status. Like the status of the live-in caregivers in British Columbia, this means that he is allowed to live temporarily in a country in order to work (see Bailey et al. 2002 for more examples).

In the novel, the protagonist is always viewed as 'other': traditional and quaint, on the margins for an array of reasons, and lost as a result. The protagonist alternately expresses longing and hatred for his home country. He spends his days translating ancient texts from a language that no longer exists in order to preserve them. He marvels at the comforts of modern European affluence – as in the shiny newness of expensive kitchens of Parisian apartments – at the same time observing its excesses, such as the mechanized, routine removal of waste. The tone of the novel expresses his alienation from the society where he lives as a temporary migrant. The novel's disorienting, non-linear narrative eventually reveals that the protagonist kills his lover, Franz, who embodies Western Europe in the very site that he loathes: the kitchen. But he continues to go unseen, with even this heinous crime unnoticed.

In many ways, Onnepalu's character embodies a contemporary theory of citizenship that has captured the attention of political geographers: Italian philosopher Giorgio Agamben's (1998) idea of the 'state of exception'. Agamben, like Hannah Arendt (1958), argues that states are able to discriminate against and violently exclude populations, including refugees, prisoners and others, from protection by the law in states of crisis. In the post 9/11 security climate in which not only terrorists but asylum seekers and other persons on the move internationally were caught up in webs of surveillance and security, the state of exception captured the imagination of political geographers (e.g. Gregory and Pred 2006).

Future research in political geography will need to contend not only with the ways that formal modes of citizenship and belonging are changing, but with the ways that these formal changes intersect with more informal practices, including the experiences of those essentially without citizenship, without formal recognition by any nation-state.

KEY POINTS

- In political geography the term 'citizenship' refers to belonging or membership in relation to territory, usually the nation-state.
- With the development of the modern nation-state, citizenship became a formalized way for states to take a census of population and ultimately categorize and monitor membership. As a result, people are increasingly dependent on states to secure a formal identity.

- The process of citizenship is often exclusionary because it creates a typology of who does and does not belong in a given territory.
- Citizenship is increasingly fluid, with people's sense of belonging crossing boundaries or encapsulating multiple territories.

FURTHER READING

Agamben, G. (1998) *Homo sacer: Sovereign Power and Bare Life*. Stanford, CA: Stanford University Press.

Ong, A. (1999) *Flexible Citizenship*. Durham, NC: Duke University Press.

Pratt, G. (2004) *Working Feminism*. Philadelphia: Temple University Press.

25 POSTCOLONIALISM

Mary Gilmartin

Definition: A Temporal Period, a Critical Approach

Blunt and Wills suggest that postcolonialism has two key meanings. The first is temporal, referring to the period after colonialism (for many former colonies, this occurred after World War II). The second is critical, referring to a range of approaches to the study and understanding of colonialism and its aftermath (Blunt and Wills 2000). Both definitions are contested.

In relation to the temporal understanding of postcolonialism, Anne McClintock argues that the use of the term post colonial is 'prematurely celebratory and obfuscatory' (1992: 91). She points out the various ways in which colonial power relations persist after formal independence: economically, politically, culturally and militarily. More recently, Jane Jacobs (1996) has commented on the fantastic optimism of the 'post' in postcolonial studies. Though less critical than McClintock, she too highlights that the legacies of colonialism make it difficult to assert that the colonial period is indeed past. These critiques are summarized in Ania Loomba's comment that 'it is more helpful to think of postcolonialism not just as coming literally after colonialism and signifying its demise, but more flexibly as the contestation of colonial domination and the legacies of colonialism' (2005: 16). In relation to the critical understanding of postcolonialism, two types of criticism emerge. The first is that postcolonial theory is more concerned with culture than with materiality. The second is that postcolonial theory is more concerned with the historical than with the contemporary (see Blunt and McEwan 2002; Gilmartin and Berg 2007).

These disagreements over the meaning of the term 'postcolonialism' point, in many ways, to the power of the concept. Robert Young, in an extensive survey of the term and its various meanings and uses, attempts to resolve these disagreements. He argues that postcolonial theory combines a 'critique of objective material conditions with detailed analysis of their subjective effects' (Young 2001: 7). For Young, postcolonialism 'commemorates not the colonial but the triumph over it' (2001: 60).

Evolution and Debate: A Matter of Representation

Postcolonialism in Geography as a Whole

The relationship between geography and studies in postcolonialism has taken many forms. Early postcolonial approaches in geography, which were not necessarily described at the time as postcolonial, focused on the aftermath of colonialism: on the unequal relationship between former colonies and colonial powers and on the nature of this relationship within a broader capitalist world system. Some geographers writing in the 1960s and 1970s were influenced by dependency theory, often as a reaction against more dominant theories of modernization (McGee 1974; Slater 1973, 1977). Power and Sidaway outline the ways in which geographic studies of dependency were highlighted in the radical journal *Antipode*, established in 1969, as well as the length of time it took for these critiques to attract more sustained engagement within geography (Power and Sidaway 2004). An expanded notion of postcolonialism developed within geography during the 1990s. Jonathan Crush, for example, articulated a vision for a postcolonial geography that would involve

> the unveiling of geographical complicity in colonial dominion over space; the character of geographical representation in colonial discourse; the de-linking of local geographical enterprise from metropolitan theory and its totalizing systems of representation; and the recovery of these hidden spaces occupied, and invested with their own meaning, by the colonial underclass. (1994, quoted in Blunt and McEwan 2002: 2)

Despite the breadth of Crush's call, postcolonialism in geography has tended to focus on issues of representation, particularly in relation to the colonial period (for more information on the concept of representation see Chapter 26). Since the early 1990s, a wide array of books has been published on this issue. This includes edited volumes on the broad relationship between geography, colonialism and imperialism, such as *Geography and Empire* (Godlewska and Smith 1994), or *Writing Women and Space: Colonial and Postcolonial Geographies* (Blunt and Rose

Figure 25.1 The General Post Office, Dublin, Ireland, a contested site in Irish colonial and postcolonial history

1994). There are also more specific texts that deal with particular aspects of a contested colonial past, such as the study of Vancouver Island by Daniel Clayton (2000), Alan Lester's work on the nineteenth-century relationship between South Africa and Britain (Lester 2001), or Yvonne Whelan's work on Dublin in Ireland (Whelan 2003) (see Figure 25.1). Anthony King (2003) has described geography's approach as the critical colonial histories approach: using colonial discourse analysis to interrogate historical archives, with the aim of drawing political and social attention to issues of representation and equity for both the colonial past and the supposedly postcolonial present. This approach has been heavily influenced by the work of Edward Said (see Figure 25.2). For many postcolonial theorists, Said is perhaps the central figure in the development of postcolonial theory. His concept of Orientalism – the imaginative geographies of the East produced by and for the West – has been inspirational across a variety of disciplines. Within geography, it has informed critical analysis of the ways in which colonialism worked

Figure 25.2 Edward Said
(Reproduced with permission from http://www.passia.org, Mahmoud Abu Rumieleh, Webmaster)

through representations of spaces and their inhabitants, as well as critical analysis of the complicity of geographers and the discipline of geography in this practice. (For more information on the Other and Othering see Chapter 28.)

While these analyses have proven a valuable addition to postcolonial studies, this is often at the expense of a broader engagement with postcolonialism that expands its temporal and substantive reach and identifies the role of patterns of injustice that developed under colonialism and inform ongoing struggles over space (King 2003). There are some exceptions, such as recent work on colonial and postcolonial cities (see Jacobs, 1996; Myers, 2003). In general, however, postcolonialism as practised in geography is more concerned with critical colonial histories. As a consequence, there have been recent calls for a rematerialization of postcolonialism in geography and for a renewed focus on the present and not just the past (Blunt and McEwan 2002: 5; Gilmartin and Berg 2007).

Postcolonialism and Political Geography

Though growing in importance in the discipline of geography, postcolonialism has a tenuous foothold within political geography. A small number of political geographers make use of Said's concept of Orientalism (see, for example, Culcasi 2006), or examine political conflict in the aftermath of colonialism (see, for example, the special issue of *Political Geography* on the Irish border: Coakley and O'Dowd 2007), but few engage in a substantive way with postcolonial theory. There are some notable exceptions (e.g. work by James Sidaway 2000; Jenny Robinson 2003a, 2003b; and Matt Sparke 2003), but in general few political geographers have taken on the challenges posed by postcolonial theory to the ways in which political geography is practised and may be reimagined. Joanne Sharp links postcolonialism to feminism, arguing that both challenge dominant forms of knowledge (Sharp 2003). However, she suggests that postcolonialism in political geography is concerned with abstract theorizing rather than having an impact on actual practices within the academy, and argues that a postcolonial political geography should embrace the specifics of empirical research in place (Sharp 2003: 69–71). Jenny Robinson makes a similar argument about the need for located research, but in a more positive vein suggests that political geography is admirably placed to engage with postcolonial theory in this manner.

In a series of articles, Robinson suggests that geography in general, and political geography in particular, could take postcolonialism seriously by extending the geographical scope of empirical studies and thus revitalizing area studies; by provincializing knowledge produced in Europe and North America that claims to be universal; by re-engaging with regional and located scholarship; and by transforming the ways in which such scholarship is produced and circulated (Robinson 2003a, 2003b). In the next section, I discuss one such recent engagement: Derek Gregory's *The Colonial Present*. I also discuss the implications of postcolonialism for our understanding of the nation-state, one of the key concepts of political geography.

Case Studies: Afghanistan and Britain

Afghanistan

Published in 2004, Derek Gregory's *The Colonial Present* represents a recent attempt to apply the principles of postcolonial theory to the ongoing War on Terror. Gregory's concern with the War on Terror is not unique: in recent years, a variety of texts have opened with or are framed by the events of 9/11 and its aftermath (see, for example, Dodds 2005; Harvey 2005; Smith 2003). *The Colonial Present* is also framed in this way: Gregory writes that the attacks on the World Trade Center and the Pentagon, and the responses to those attacks, forced him to pay attention to the contemporary resonance of the ideas and practices that previously he had studied only in a historic context. Gregory also highlights his debt to Edward Said and the concept of Orientalism, as well as to the essays of Arundhati Roy. In the body of the book, Gregory seeks to show that the 'War on Terror' is 'a series of spatial stories that take place in other parts of the world' (Gregory 2004: 13). Though this discussion focuses on Afghanistan, the book also considers Iraq and Palestine; all parts of the world that were not only shaped by their colonial past but continue to be shaped by their colonial presents.

Afghanistan has long been a site of struggle between colonial and imperial powers. Britain fought three Afghan wars in the nineteenth and early twentieth centuries in order to secure the borders of the Indian Raj and to ensure control of the Khyber Pass (Gregory 2004: 30). The border between Afghanistan and British India (later Pakistan) was drawn by the British colonial administration in 1894–95, dividing the territory of the Pashtun, one of the main ethnic groups in Afghanistan (see Figure 25.3). In addition to the Afghan wars, Britain and Russia were also involved in a struggle for influence over the territory, a struggle that led, in the early twentieth century, to the formation of Afghanistan as a modern state out of a range of rival fiefdoms (Gregory 2004: 31).

Afghanistan was thus originally a site of conflict between Britain and Russia. Later, it became a site for the struggles of other imperial powers when, during the Cold War, both the USA and the USSR provided it with significant levels of aid. The USSR invested particularly heavily: between 1956 and 1978, it provided $2.5 billion in economic and military aid to Afghanistan (Gregory 2004: 33). The USSR also invaded Afghanistan in 1979 in order to reinstall a Marxist-Leninist government. In the civil war that followed, the government was supported by the USSR, while the resistance movement – the mujahideen – received support from the USA, Pakistan and Saudi Arabia (Gregory 2004: 34–36). The USSR finally withdrew all its troops in 1989, and the US also withdrew its support (Gregory 2004: 37). However, the legacy of the conflict remains. One of the mujahideen was Osama bin Laden; others became involved in the Taliban, which declared Afghanistan an

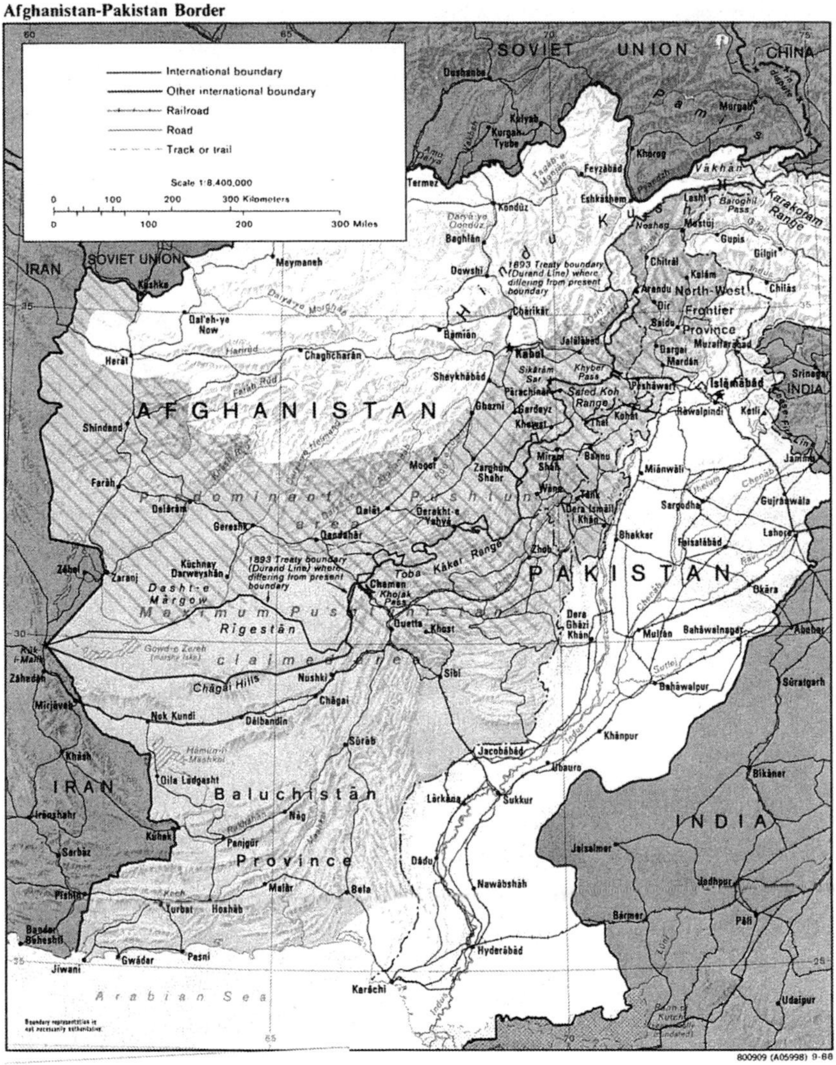

Figure 25.3 Map of the Afghanistan-Pakistan border. The shaded area shows places inhabited by the Pashtun ethnic group on both sides of the border, 1988. (Courtesy of the University of Texas Perry-Castañeda Library Map Collection)

Islamic Emirate in 1996. The Taliban were removed from power following the US-led invasion of Afghanistan, which took place in October 2001 as a response to the 9/11, 2001 attacks.

In this way, Gregory shows how Afghanistan became a stage on which western conflicts were played out, in a way that constructed Afghanistan as a spatial and social other for the purposes of western imperial fantasies. This process of 'othering', Gregory points out, has material consequences, not least in the ways in which 'Afghanistan was transformed by the role it was recruited to play' (2004: 45). However, Gregory also insists that contemporary Afghanistan is not just a product of conflicts between superpowers, but also involves the active involvement of other individuals and organizations. Despite this, the powerful discursive construction of Afghanistan for the purposes of the War on Terror, as a place of barbarism that harboured international terrorists such as Osama bin Laden, shows the ongoing relevance of Orientalism, as well as the 'destructive power ' of its imaginative geographies (Gregory 2004: 75). He concludes *The Colonial Present* with a call for a reflection on both history and geography, on the 'multiple ways in which difference is folded into distance', and for a realization that distant strangers are neither all that different nor all that strange (Gregory 2004: 261–262). Gregory draws on colonial histories and postcolonial theory to illuminate the processes at work in a contemporary world at conflict. His book offers an excellent antidote to those who critique postcolonialism for its lack of contemporary relevance.

British Muslim Women and the Meaning of 'Home'

Another way in which postcolonialism matters to political geography is through its questioning of the contemporary nation-state. In the introduction to their edited volume on *Postcolonial Geographies*, Alison Blunt and Cheryl McEwan comment on the predominance of chapters dealing with cultural and/or historical research. That said, some of the chapters offer considerable insights for political geographers through their discussion of belonging. This case study focuses on one of those chapters, Claire Dwyer's discussion of the meaning of 'home' for young British Muslim women (Dwyer 2002). Dwyer argues that the young British South Asian Muslim women she interviewed occupy postcolonial spaces, spaces that were shaped by 'a particular set of postcolonial migrations set in motion by the operation of social, political and economic interlinkages between the metropolitan core and the (post)colonies' (2002: 184). Their presence in Britain, and the complicated ways in which they negotiate that presence and their relationship with South Asia, highlight the ways in which national spaces and national identities are fundamentally challenged or altered as a consequence of postcolonialism. As one of the young women Dwyer interviewed said: 'I identify not as a British or an Asian but as a British Asian. I'm an Asian, I've got an Asian background but I've been brought up in British society' (Eram, cited in Dwyer 2004: 188). Dwyer describes this as 'the shaping of a postcolonial British identity' (2004: 198). She draws on the work of Stuart Hall (1996) to suggest that the postcolonial is primarily concerned with decentring a grand narrative based on the existence and pre-eminence of the nation as a form of territorial and ideological construct.

KEY POINTS

- Postcolonialism has two key meanings: temporal (the period after colonialism); and critical (the study of colonialism and its aftermath).
- Within geography, postcolonial approaches have generally focused on questions of representation, particularly in relation to the colonial period. This work has been heavily influenced by Edward Said.
- Political geography has been slower to engage with postcolonial theory, but recent work has used postcolonial theory to gain insights into contemporary political geographies.

FURTHER READING

Blunt, A. & McEwan, C. (eds) (2002) *Postcolonial Geographies*. London and New York: Continuum.

Blunt, A. & Wills, J. (2000) *Dissident Geographies: An Introduction to Radical Ideas and Practice*. Harlow: Pearson Education.

Gregory, D. (2004) *The Colonial Present: Afghanistan, Palestine, Iraq*. Malden, MA and Oxford: Blackwell.

Said, E. (1979) *Orientalism*. New York: Vintage.

Sidaway, J. (2000) 'Postcolonial geographies: an exploratory essay', *Progress in Human Geography*, 24(4): 591–612.

26 REPRESENTATION

Peter Shirlow

Definition: The Production of Meaning

Representation constitutes the manner through which ideas, beliefs, values and images are both produced and provided with meaning. We are surrounded by all sorts of representations of space and place from television, advertising, graffiti, walls murals, speech, music, newspapers, paintings and photographs. None of these mediums are value-free; they communicate, often intentionally, an idea, the synthesis of ideas and/or the nature of power relationships. Text and image are in effect the constitution of what people consider to be reality or, more importantly, the imagination of reality (Barnett 1997).

In many instances official (i.e. state-based) representation aims to displace sections of the population from narratives of place. Official discourses of representation (state-sponsored parades; national monuments) highlight the ground upon which the dominant discourses of a society are both imagined and managed. Non-official representations (graffiti; murals) pinpoint a desire to include the excluded (women; ethnic minorities) who have become invisible within dominant forms of representation. Ultimately, the representation of ideas is always produced by and for a specific group. The key questions asked by those who study representation concern the motives and authority of the presenter (Pred 1997). Thus representation is based upon the 'self' and the 'other' and the reflexive practices that constitute such division.

Evolution and Debate: The Role of the Cultural Turn

Within political geography the concept of representation has traditionally been considered quite narrowly as the mechanism by which political representatives are chosen to govern a given political unit. After the cultural turn, however, representation has taken on another meaning altogether.

The Cultural Turn

The 'cultural turn' in geography was based upon a rejection/modification of economic determination and a need to locate representation within a cultural context. The emerging of new cultural geographies, post-structuralism, post-colonialism and post-feminist studies rejected more structured explanations of representation and in so doing aimed to understand more about how visual images, language and the communication of ideas influenced the constitution of 'reality'. Despite varied approaches to theories of representation the shared aim of these approaches is to encourage an understanding and appreciation of the relationships and links, whether formal or informal, between discourse, value, gender, class, ethnicity, culture, politics and all other relationships that constitute the production and reproduction of image, text, landscape and place (Rose 2001). The interpretation of representation is always linked to evaluating the relationship between the social and the cultural and the impact of time and space upon them. The study of representation is an analysis of process (Whatmore 2002).

The challenge to the authority and rationale of official representations has led to the deconstruction of various dominant forms of representation. This has involved a series of approaches including challenging masculine-centred renditions of geography, the authority of dominant cultures to speak for all cultural groups and the demotion of once dominant essentialist theories of what constituted place. The 'cultural turn' aimed to ensure that geographers when studying representation located the excluded and the marginalized and in so doing mobilized less unitary and exaggerated notions of place in favour of more nuanced and pluralized spaces (Castells 1996; Thrift, 1996). Evidently, this has meant that geographies of representation are now understood as geographies of place that are more complex and intersecting (Barnett 1997). These shifts in geography and the wider appreciation of hybridity, third spaces and heterogeneity within place as well as the contest between dominant and resistant discourses further the notion that all forms of representation are differentiated not only by the positions of the presenter and/or the viewer but also by the relationship between knowledge and place. The processes through which geographers reinterpret the presentation of ideas and their relationship to place leads to the rearticulation of those ideas and a challenge to previous constructions of meaning. This can take place on all scales, from the global to the local. Identity issues are thus embodied in the representation of place and via complex and contested interpretations of what that representation means. As Borja and Castells write:

> The creation and development in our societies of systems of meaning increasingly arises around identities, expressed in fundamental terms. Identities that are national, territorial, regional, ethnic, religious, sex-based, and finally personal identities – the self as the irreducible identity. (1997: 13)

Why is the Study of Representation Important?

Place, space and human landscape are never blank or merely natural; they are riddled with meaning, symbolism, contradictions, various layers of history, social and cultural practices and complex power relationships. All human-spatial relationships are indicative of underlying forces, structures and imaginations that drive the perception and presentation of order, belief and ultimately our attitude to place. Each mode of representation, and the communication of meaning from it, is ultimately contested given that the presentation of place determines the nature of control and the rejection of being controlled. In effect place affects discourses of representation and vice versa (Latham and McCormack 2004).

Representation is a powerful medium through which actors and agents reproduce and enhance preferred images, which they present as accurate and entirely 'truthful'. In essence, all representation is based upon a propaganda conditioning perspective which aims to subvert the reader. This does not mean that all forms of representation are malign, devious or untrustworthy. However, we should be reminded that the presentation of any belief or idea is based on the views of those making the representation and is therefore a rejection of ideas and actions that do not synthesize with them (Rose 1993).

The work of Cresswell (1996) has been central in reminding us that the representation of ideas is usually accepted uncritically because many social and cultural relationships are taken for granted. Thus the study of representation is not merely centred upon interpreting and studying the meaning of the text, image or ideas presented or simply concerned within understanding the geography of rejection and resistance to universalizing representations, but is also concerned with determining why imagined, exaggerated and hostile representations are not challenged. Thrift (1996) has argued that studying those who do not challenge what is presented to them is part of the analysis required to discover why such an apolitical approach is undertaken. Evidently, the geographies of power and representation involve not only the presentation and rejection of ideas and values but also complicity with them. This reminds us of the dilemma faced by geographers who seek to use various academic arguments to awaken and reposition the acceptance of 'truth, 'normality' and the representation of both (Brinegar 2001).

We can consider these complexities of representation when we think of the growth in digital technology and the production of goods which allow us to play games, listen to music and download videos. Such technologies are presented as the 'must haves' of a generation. They are represented via advertising as sophisticated pleasure-giving instruments. But if we consider the production of such goods we will find that their meaning may be different for the worker in the industrializing world who manufactures them. Representation may shift from the presentation of fun to concentration upon low-paid work, un-unionized workplaces and even the use of child labour. The representation of the child's hands playing with such technology in the 'advanced' world is somewhat different to representation of the child's hands assembling these 'instruments of pleasure' in squalid working conditions. Campaigns by groups that have represented the oppressed have highlighted this dichotomy but such efforts do not mean that affluent Westerners have abstained from buying such goods. Does this mean that consumers do not care? Is it that the power of advertising is so immense that it subverts decency? Does giving to charity offset any sense of guilt concerning global capitalism? Are the pressures of children upon parents to buy such goods important sites for exploration? Whatever the answer, the variant representations of the consumption of goods produced via unethical labour practices stretch across a series of global, national, societal community and home boundaries.

Understanding that representation is never representative of everyone and 'every place' is a key concern of geographers. Every place is represented via a perspective of including and excluding and also feeling excluded. Nearly every urban centre has a city hall that acts as a site of commemoration and civic pride. Statues, stained-glass windows and paintings illustrate a version of history and chosen historical events. These sites usually commemorate 'key' industrialists, the 'dead' of wars and 'important' civic leaders – most of whom are men. The role played by women and ethnic minorities in such key events is generally omitted. Generations have grown up with the Western movie genre in which Native Americans are cast as violent and dangerous. Media-driven representations persuade us that certain places, areas of social exclusion, are dangerous and to be feared. In recasting these forms of representation we can reposition those who are invisible in such renditions of place and community and formulate more reasonable depictions. Addressing such exclusion is vital but so too is an appreciation that the casting of such unrepresentative sites has been part of the overall process of labelling space, excluding powerless minorities and removing the viewer from radical and inclusive conceptions of place.

In resisting dominant discourses and the strictures of society the act and performance of anti-social behaviour can also be representational. The creation of graffiti and unofficial murals is a key act in reclaiming a sense of ownership over place. In challenging official representations of place it is evident that rioting, the mobilization of street protest, the undermining of police and thus state authority are not merely geographies of frustration and malice but instead a reaction to exclusion and in some instances the expression of a desire to be acknowledged and heard. The activities of street gangs, who are often motivated by a desire to claim territory, and their sense of dress, violent enactment and linguistic performance symbolize both resistance and an attempt to depict their interpretation of place and their representational position within it.

Given such complexities in understanding what representation constitutes and how we study, it it is important to position methodological approaches. These include:

- Who is the presenter?
- Why are they making these representations and what is their motivation?
- What does the presenter want us to see and/or understand from their representations?
- Who is their audience?
- How is the representation composed?
- Is there a hidden text or sub-theme in what is being presented?
- Who would accept these representations and why?
- Who would reject these representations and why?
- Who is included/excluded by the representations offered?
- What social, cultural historical and other relationships are being invoked/excluded?

- How might these presentations influence place and the interpretation of it?
- Is more than one interpretation of what is being represented feasible and not contradictory?
- What knowledge is required to deconstruct representations?
- How does this representation affect the construction of space and place?
- What as a geographer should I do if I find the representation offensive?

Case Study: Two Stories on a Wall

Northern Ireland recently endured a thirty-year civil war, known euphemistically as the Troubles. The war ended in 1998 with a peace accord signed on Good Friday in the city of Belfast. The Troubles is synonymous with a territorial dispute between those who want a United Ireland (mostly Catholics who ascribe to Irish nationalism or support the Irish Republican Army) and those who wish to remain in the United Kingdom (mostly Protestants who support Unionism or loyalist paramilitary groups). Since the late 1960s over 3,600 people have been killed in Northern Ireland and many more were injured or psychologically scarred. This violence produced and intensified the segregation between Protestants and Catholics (Boal 1996).

Such residential segregation is usually linked to wider processes of territorial marking, including murals, flags and kerb painting. Although the contestation over space and territorial control is no longer achieved through the high levels of violence common up to the mid-1990s, it is still evident that the commemoration of identity aims to provide some certainty in the context of change that has accompanied the Irish peace process. The use of murals, in particular, is an articulate form of representational symbolism and an illustration of territorialized power. The events and images displayed in these murals tend to express 'dominant' views of territorial belonging and community solidarity in the face of conflict. The marking of space in such circumstances pinpoints selection by reminding the viewer of oppression and the need to celebrate examples of armed and civil resistance to that oppression. Such markings stretch across a spectrum from what Graham and Shirlow (2001) identify as the mnemonic to the reinforcement of cultural apartness.

The representation of spatially proximate events, that will be understood by those who live in the areas where murals are painted, positions the link between place and experience. Muralists use distinct spaces within which to promote a propaganda conditioning perspective, which encompasses signals of territorial demarcation (Shirlow and Murtagh 2006). Spatially sensitive experiences have strong symbolic codes and messages attached to them. This process can be seen in relation to two murals – one produced by the nationalist/republican side and the other produced by the unionist/loyalist side.

Ardoyne and the Holy Cross Dispute

The Holy Cross dispute occurred between 2001 and 2002 in the Upper Ardoyne area of Belfast and involved escalating tensions between loyalists and nationalists/republicans. The dispute centred upon a Catholic girls' primary school located within a loyalist area. Loyalists picketed the school and used offensive and abusive tactics against schoolchildren and their parents as they walked to school. Loyalists claimed they were responding to Catholic parents using the school run to cause damage and harass the local Protestant community. Republicans and nationalists (most of whom lived in Ardoyne) countered that the picket was an infringement on the rights of parents and their children to walk to school free from sectarian harassment.

In the wake of the protest a mural was painted that commemorated the events around the Holy Cross dispute and the impact it had upon residents within the Ardoyne community (see Figure 26.1). The mural captured the sense of fear that the schoolgirls had to endure. The central panel depicted the vulnerability of women taking small girls to school along a road of 'loyalist bigotry'. The use of the symbol of a mother and her depiction in shadow foregrounds the children as the most vulnerable of all.

The mural also established a visual connection/relationship between the Ardoyne struggle and one in Little Rock, Arkansas, in the United States during the height of the American civil rights movement. The Little Rock incident is often called the 'Crisis at Central High' and centred on state-led efforts to desegregate the city's schools. Nine African-American pupils enrolled at school were debarred by racist protests in the area and were only forcibly enrolled after President Eisenhower dispatched federal troops to ensure the students' safety and enforce their right to attend. There are two reasons for the inclusion of this event from 1957 in the Ardoyne mural. First, the creation of an international lineage of suffering unites the local community with wider struggles for rights and recognition. Second, the events in Little Rock are used to

Figure 26.1 Mural depicting the Catholic view of the Holy Cross Dispute in Ardoyne, Belfast, Northern Ireland. (Photo by and courtesy of Peter Shirlow)

represent differing state actions regarding both events. The mobilization of the federal troops to ensure that schooling in Arkansas was desegregated contrasts with the police officer on the panel on the right standing emotionless despite the obvious intimidation of fearful children.

However, evaluating such representations is important and indeed we should seek to challenge part of the interpretation presented as the issues that related to Arkansas were about desegregation whereas the parents of Holy Cross sent their children to a Catholic school and were not seeking the desegregation of the local schooling system. In essence the relationship represented between the two events was somewhat erroneous, although the obvious abuse suffered was exact and true.

Shankill Seeks Justice

The second mural also plays upon the idea of victimhood; however, it is produced by and for a predominantly loyalist/unionist area known

Figure 26.2 Mural highlighting Republican attacks in the Protestant Shankill Road area of Belfast, Northern Ireland
(Photo by and courtesy of Peter Shirlow)

as the Shankill Road (see Figure 26.2). In many ways this mural is a direct representational response to the Ardoyne mural discussed above. Indeed, the mural calls attention to several bombing attacks on the Shankill Road by Irish republicans, notably some who came from the Ardoyne area. At the top of the mural is a direct challenge to republican representations of themselves and their actions as non-sectarian. The panels depict images from newspapers at the time of chosen bombing incidents, along with their location and number of victims killed. The word 'innocent' is placed before the list of victims, playing upon the evocativeness of civilian deaths and especially, as shown in the central panel, the murder of children. The allegation of republican sectarianism is also emphasized in the mural by reminders that these sites of violence were not military, economic or typically 'legitimate' targets. In effect the Shankill community is represented as transgressed against and victimized. Unlike the Holy Cross mural where police are presented as unhelpful, if not complicit in the abuse, the Shankill mural presents police, ambulance and fire brigades, and more crucially the army as aiding those who suffered. In the two murals the negative and positive casting of the police conveys the differing attitude to these organizations by the respective communities. In analysing this mural it is clear that the uncomfortable realities of violence propagated by members of the Shankill community against Catholics, the state and Protestants from that community is omitted. In sum both murals present unidimensional and selective narratives of suffering.

Each of these representations highlights opposition towards 'other' groups via 'grounded' and 'lived experiences'. Representation is driven by subjective interconnections and emotional forces that drive community-based 'morality'. These constantly negotiated and contested social and spatial practices matter in that they are interpreted and given significance by their participants. They refer to the capacity of social groups to develop distinct patterns of separated life. Thus, the need remains for those who study representation to locate the structural relations of power, displacement and conflict, which pinpoint how spaces are both (re)constructed and contested.

Evidently such representational efforts do not promote pluralist or shared interpretations, as they intend to promote wider narratives of separation. Within an intensely divided society it is evident that mobilizing harm, via commemoration, murals and other sites of representation, is undertaken through recognition of the harm endured and, more importantly, acknowledged within selective memories. Representation remains exclusive and controlled.

KEY POINTS

- The study of representation has shifted beyond a mere analysis of government and territorially defined political units.
- Representation is linked to studying the formal and informal relationships between discourse, text and imagination.
- Place and the human landscape are neither blank nor neutral but laden with meaning, symbolism and contradiction.

FURTHER READING

Graham B. and Shirlow P. (2002) 'The Battle of the Somme in Ulster memory and identity', *Political Geography*, 21: 881–904.

Latham, A. and McCormack, D. (2004) 'Moving cities: rethinking the materialities of urban geographies', *Progress in Human Geography*, 28: 701–724.

Whatmore, S. (2002) *Hybrid Geographies: Natures, Cultures, Spaces*. London: Sage.

27 GENDER

Alison Mountz

Definition: Gender as Social Construct

Gender is more a process than a 'thing'; it refers to the processes wherein people are assigned qualities on the basis of their sex. Sex has traditionally been separated into binary categories – male or female – that are associated with biological features. By contrast, the social attributes commonly assigned to sex categories demonstrate that gender is a *social construct*. In other words, there is no underlying, essential quality attached to what it is to be 'male' or 'female'; instead attributes, features and behaviours are assigned or deemed appropriate in relation to men and women. Gendered interactions are structured by social relations and vary across time and space.

These attributes are then transposed on to and into the ways that people interact: behaviours deemed 'appropriate'. As Linda McDowell (1999: 7) suggests, 'what people believe to be appropriate behavior and actions by men and women reflect and affect what they imagine a man or woman to be and how they expect men and women to behave, albeit men and women who are differentiated by age, class, race or sexuality, and these expectations and beliefs change over time and between places'. Men and women are valued differentially in different locations and economic spheres. Men overpopulate corporate offices where women are underpaid relative to male counterparts; women overpopulate elementary school classrooms where they are considered to be nurturing and naturally good with children. Men work in construction; women in nursing. Within the home, women are made to feel more comfortable in the kitchen and men in the masculine spaces of basements or garages.

Gender categories have increasingly come under fire in recent years for the exclusions they invoke discursively and materially. Judith Butler (2004) argues that these and other binary categories enact violence. For intersexed persons, for example, who have biological components of both male and female sexes, exclusion from the ability to inhabit either category enacts violence. The policing of the categories of man and woman excludes persons who do not fit clearly into either. They may be excluded, for example, from single-sex institutions and spaces as elite as private women's colleges (see Quart 2008) and as mundane – though by no means banal – as public bathrooms.

Evolution and Debate

Feminist Geographers on Gender

Gender and geography have always been connected (e.g. Domosh and Seager 2001; McDowell 1999; Massey 1994). Feminist geographers have looked at a broad range of sites and scales to understand how gendered relationships unfold in specific places and in turn shape those places, crafting gendered architectures and landscapes that reflect power relations among people.

The history of feminist geography and its attention to gender in many ways reflects the broader feminist movement. Feminist geography began with the feminist social movements in the 1960s and 1970s with the goal of including women in equitable ways: in society, politics, the workforce and education, to name but a few realms. Universities, for example, were shamed for not hiring, promoting or otherwise valuing women. The discipline of geography, too, was admonished for not conducting research that included 'half of the human race' (Monk and Hanson 1982) and for not valuing, researching or promoting women and feminism within the discipline (Rose 1993).

In addition to institutional changes, feminist scholarship sought, quite simply, more inclusion of women in research. In geography, this meant an interest in researching the experiences of both men and women. Feminist scholars aimed to include the experiences of both men and women in labour markets (Hanson and Pratt 1995), access to health care (Dyck et al. 2001) and public transit (McLafferty and Preston 1991).

Early feminist research by scholars in western countries was critiqued and challenged for focusing primarily on the experiences of white, middle-class women (Mohanty 2003). Feminists responded by theorizing difference more broadly and more inclusively, with attention to the influence not only of gender, but of class (Saltmarsh 2001), race (Kobayashi and Peake 1994), sexuality (Bell and Valentine 1995) and ability (Chouinard and Crooks 2005). Such analyses of the intersecting axes of difference were called 'intersectional' analyses (e.g. Valentine 2007).

Eventually, and always in line with feminisms across the humanities and social sciences, feminist geographers moved from the theorizing of difference to identity more broadly (e.g. Alcoff 2006). Consistent across these conceptual shifts has been a sustained interest in how space and place shape and in turn are shaped by gendered processes. Does a sexist society shape a gendered workplace, or does a sexist workplace cause gendered relations? Most likely, the answer is both. Recursive processes occur at multiple scales and shape how we interact with one another.

The ways that gender is framed by and frames spatial relationships have always been of interest to feminist geographers. In a classic study conducted in the 1990s in Worcester, Massachusetts, Susan Hanson and Geraldine Pratt (1995) researched how far women travelled from home to participate in the labour market. They found that in spite of great differences among the households surveyed, whether or not they had families and regardless of the kind of work they did, women tended to work closer to home than men. This finding confirmed beliefs that women were more readily associated with and responsible for work at home.

While some feminist geographers focused on life in cities, still others looked at multiple scales. Doreen Massey (1994), for example, studied how identities were constituted through social and economic processes that were simultaneously local and global. More recently, transnational feminisms (e.g. Sangtin Writers Collective and Nagar 2006) have looked at how social movements intersect and interact across national borders.

Sex and gender are no longer the stable categories that they were once presumed to be. As intersexed persons demonstrate, some people inhabit multiple categories of sex, rather than fitting neatly into one or the other (see Figure 27.1). Poststructuralist theorists, Judith Butler (2004) most notable among them, have argued that there are no underlying or 'essential' qualities that distinguish men and women. Rather, we produce and assign qualities to gender categories through social interactions and discourses. Boys are taught to be tough and not to cry, and girls are socialized to be nice and play well with others, to offer some common examples of gendered assumptions.

This destabilization of the very categories of sex and gender has posed challenges to feminism and feminist movements. If the category

Figure 27.1 Symbol for transgender/intersexed people

'woman' is one that excludes so many people, how can it serve as the basis of organizing for feminist social movements? Yet in spite of these challenges to feminism, feminist research presses on. Some people look at the spaces of home (e.g. Dowling and Blunt 2006) to understand how men and women occupy spaces of dwelling distinctly, and how these dwellings reflect gender relations. Others examine interactions and spatial distributions in workplaces (e.g. Wright 2006). They ask why women workers are concentrated in particular occupations, how they negotiate power relations and why they still earn less than men doing the same job. Still others conduct analyses at a global scale, examining the international division of labour (e.g. Mullings 1999).

Some of this research takes categories of sex and gender for granted, while some seeks to understand how the ways that people live their lives destabilize gendered hierarchies, challenging the very categories critiqued by Butler (2004). Intersexed persons, for example, occupy multiple categories at once, creating new gender categories. Transgendered persons also move between categories of gender, sometimes taking up multiple positionings through physical embodiment and social ways of being (Feinberg 1998).

Political Geographers and Gender

While political geography as a subdiscipline has not always embraced feminism or the study of gender (see Desbiens et al. 2004), feminist research has always advanced understandings of the topics important to political geographers, whether feminist theories of the state, war or nationalism. Feminist political geography followed a similar trajectory to feminist research throughout the discipline, beginning with an interest in including women in understandings of electoral processes and politics (e.g. Secor 2004) and evolving to understand how power and political processes shape and in turn are shaped by gendered relations (e.g. Cope 2004).

In 2004, Lynn Staeheli, Eleonore Kofman and Linda Peake published an important book, *Mapping Women, Making Politics,* which offered 'a step in demonstrating the ways in which feminist perspectives on politics and political geography contribute to better, richer understanding of political processes, activities, and behaviors' (2004: 1). The editors outlined an explicitly feminist research agenda within political geography, one comprised of 'situated knowledges that are derived from the lives and experiences of women in different social and geographic locations' (2004: 1–2). In order to do so, they include analyses of different kinds of political activity from very different spheres. The authors collectively address the more formal sorts of political engagement such as voting practices as well as the more violent and forced, if equally political, such as rape as a tool of violence in war (Mayer 2004), and displacement and mobility as explicitly politicized acts (Hyndman 2004). In this way, Staeheli, Kofman, and Peake emphasize the importance, if not the centrality, of so-called 'small p' politics to spatial organization. (See the Introduction to this volume for more information on 'small p' politics in political geography.)

In the next section two case studies drawn from this edited collection – one on gendering politics, the other on gender and war – are detailed. They reflect roughly the trajectory of attention to gender by political geographers.

Case Studies

Case Study 1: Gendering Politics

In her contribution to *Mapping Women, Making Politics*, Meghan Cope (2004: 71) seeks to 'continue the feminist project of expanding what counts as politics'. She argues that a gendering of political realms and relations should include not only formal aspects of political participation by women (in public spheres of politics and electoral processes and formal forms of exclusion and discrimination by the state), but also their daily engagements with life, whether small or large 'acts of resistance'. She posits three strategies to examine 'diverse political acts performed by women': women's creation of new spaces for political action, women's use of 'embedded codes of specific places' to challenge oppression, and women's efforts to 'jump scale' in their political actions (2004: 71).

Cope's essay adeptly addresses the broad range of engagements that feminists consider to be 'political', engaging the 'multiplicity, diversity, and complexity of political acts' (2004: 72). She accomplishes this goal by asking questions in multiple sites and at multiple scales. Cope begins with more formal sites and modes, namely participation in political systems and voting. She shows that it matters not only *who* participates in political acts, but *how* and *where* they do so. Women organize from the home, kitchens, schools and work sites, in order to influence more public political realms. Cope highlights the ways that women have moved private matters, such as domestic violence and disappearance, into public spaces, as in *Las Madres de La Plaza de Mayo* who organized collective demonstrations to protest about the disappearance of family members arrranged by the Argentinian government (Figure 27.2). This group of older, Catholic mothers mobilized their identities *as mothers* effectively to call attention to state violence (Radcliffe and Westwood 1993). Cope argues that the Mothers' movement demonstrates all three of the strategies she sets forth. They crafted a new site for politics by gathering to demonstrate in the city's central plaza. They turned the coding of the plaza as a site of state power into one of resistance, and they jumped scale by attracting the attention of international media (2004: 76).

Case Study 2: Gender and War

Political geographers have also looked at how women are affected disproportionately by armed conflict. Both Jennifer Hyndman (2000) and Tamar Mayer (2004) demonstrate that gender operates through nationalism with the violent use of women's bodies as tools of ethnic cleansing to advance the nation.

Hyndman (2001) addresses the documentation of rape as a tool used against women during war. She notes that an estimated 25,000 women were raped in Bosnia-Herzegovina and an estimated 250,000 women were raped in the genocide in Rwanda. Mayer also analyses the

Figure 27.2 Photo collage of disappeared people during Argentina's dirty war. *Las Madres de la Plaza de Mayo* organized collective demonstrations to protest at the disappearance of family members. *Las Madres* were able to call attention to state violence by mobilizing their identities *as mothers*. (Photo by and courtesy of Pepe Robles)

Bosnia-Herzegovina case of rape as a form of 'ethnic genocide' of Muslim women by Serb men, Croat women by Croats and Serbs and Serb women by Muslims (2004: 159). She argues that rape was 'carried out by men as a way to control, terrorize, and dehumanize women' (2004: 158), creating broader 'landscapes of fear' through which women move in their daily lives. Such fear is intensified during war. These acts took place in private homes, in public spaces, in camps and in brothels. Both Mayer and Hyndman partake in political geography that takes gender as a central subject of analysis in order to map overlapping archipelagos of violence and locate women's bodies as a significant battlefield in armed conflict. Both suggest that accounts of this rise in the documentation of violence against women resulted in the designation of rape as a weapon of war (Hyndman 2001) and eventually 'provided the basis for establishing the first international war crime tribunal after Nuremberg' (Mayer 2004: 159).

Gendered analyses of political geography such as these show the intersection of private and public lives in formations of political acts undertaken by both men (to advance the nation and nationalism) and women (to collectively protest against violence), jumping scale from home to nation to international venues to address the horrors of war.

Future Research: Bridging Gaps and Destabilizing Categories

Recently, feminist scholars have grown very interested in research on geopolitics and gender (e.g. Hyndman 2004; Sharpe 2004). They have addressed the failure of political geography to pay attention to gender (Staeheli et al. 2004).

With time, some of their concerns have been taken up by 'mainstream' political geography and they have proven effective at 'bridging the gap' between political and feminist geographers identified by Jennifer Hyndman (2004). Feminist theories of gender now inform not only the topics that political geographers study, but the methods they use in their research (e.g. Sharpe 2004), and in particular the ways that methods can address and even destabilize the differential power relations between 'researcher' and 'researched'.

Not enough research has been done, however, to destabilize the very categories of gender highlighted by writers like Judith Butler and Leslie Feinberg, and this poses the next challenge for feminist political geographers.

Furthermore, while feminist political geographers have done their part to 'bridge the gap' a good number of political geographers have not necessarily answered the call and much of political geography remains a masculinist endeavour (Dowler and Sharpe 2001; Staeheli et al. 2004).

KEY POINTS

- Sex categories – male or female – are associated with biological features. Social attributes assigned to people on the basis of their sex form a person's gender.
- Gender is a *social construct*.
- In geography feminist scholars were the first scholars to systematically study the role of gender in spatial relationships.
- In political geography, feminists have developed feminist theories of classic geopolitical concepts, including the state, war and nationalism.
- Feminists have also emphasized the importance, if not centrality, of so-called 'small p' politics to spatial organization.

FURTHER READING

Domosh, M. and Seager, J. (2001) *Putting Women in Place: Feminist Geographers Make Sense of the World.* New York and London: Guilford Press.

Massey, D. (1994) *Space, Place, and Gender.* London and New York: Routledge.

McDowell, L. (1999) *Gender, Identity and Place: Understanding Feminist Geographies.* Minneapolis: University of Minnesota Press.

Staeheli, L., Kofman, E. and Peake, L (eds) (2004) *Mapping Women, Making Politics: Feminist Perspectives on Political Geography.* London and New York: Routledge.

28 THE OTHER

Alison Mountz

Definition: A Verb and a Noun

The term 'other' serves as both a noun and a verb. By placing one's self at the centre, the 'other' always constitutes the outside, the person who is different. As a noun, therefore, the other is a person or group of people who are different from oneself. As a verb, other means to distinguish, label, categorize, name, identify, place and exclude those who do not fit a societal norm. In geographic terms to other means to locate a person or group of persons outside of the centre, on the margins. 'Othering' is the process that makes the other. 'Othering' is the work of persons who discriminate, and it has also been the work of social scientists and philosophers.

The process of creating the 'other' wherein persons or groups are labelled as deviant or non-normative happens through the constant repetition of characteristics about a group of people who are distinguished from the norm in some way. To assume that unions between men and women are the norm, for example, is to other same-sex couples. To take the census in a way that places Caucasian at the centre or single racial categories as the norm is to other 'mixed race' identities or non-white identities that complicate the checking of a singular box on a form. Such racialization places whiteness at the centre, and persons of colour on the outside of normal. To require students to pay high tuition fees and to purchase expensive equipment in school is to other those who do not have the material means to do so; it is to place those without money outside of normal. To create school curricula exclusively in one language in a bilingual setting is to 'other' those who speak any language other than that chosen; it is to place them outside of normal, to alienate them by virtue of the curriculum. To assume that every person will climb stairs to enter a building is to 'other' and exclude from entering those who move through the world in a wheelchair. Not to provide health coverage for pregnant women because pregnancy is considered a pre-existing condition is to 'other' through gendered processes; to place men at the centre of expectations for and rights to health care.

Evolution and Debate: A Rich Intellectual Lineage

A number of scholars in recent decades developed the notion of the 'other' as an epistemological concept in social theory. Postcolonial scholars, for example, demonstrated that colonizing powers narrated an 'other' whom they set out to save, dominate, control, civilize, and/or extract resources through colonization. In 1978, Edward Said published an important book titled *Orientalism*. Said showed that western powers, philosophers, artists, social scientists and many additional parties had othered 'the Orient' as a region (Figure 28.1). The process of othering relies on broadly drawn dichotomies. In the texts studied by Said, the 'self' was the western author, who set his or her sights on the East as an object of study, thus creating a binary between West and East, self and other. Said labelled this process 'Orientalism', wherein western subjects produced the East, relying on this binary to homogenize (suggesting all were the same), feminize (suggesting they were the lesser of the two), and essentialize (suggesting in a reductive fashion that the region had underlying characteristics) that scripted the East as not western – as traditional, underdeveloped, decadent, greedy and so on (Figure 28.2). Said showed how the creation and distribution of knowledge *about* the Orient created an essentialized version of the East that not only justified western colonization of the area, but also helped the West define itself as a superior, civilizing force.

In 1988 Chandra Mohanty (2003) levelled a parallel critique at western feminisms and feminists with her now famous essay 'Under western eyes'. Just as Said had suggested an 'othering' of the East by the West, Mohanty illustrated that feminists of the global north had 'othered' feminists and feminisms of the global south. They did so, she suggested, especially by the creation and reiteration of the category 'Third World women', producing knowledge *about* women in the global South, rather than learning from their diverse standpoints and struggles. Western feminists produced a stereotype of woman of the global South as disempowered, indigenous and unable to speak for herself. Mohanty argues that of course women in any part of the world cannot be placed in one homogeneous category. Women are diverse, and the category of 'woman' is therefore an unstable one (like any other category). It follows that there exists no one feminism, but multiple feminisms across which

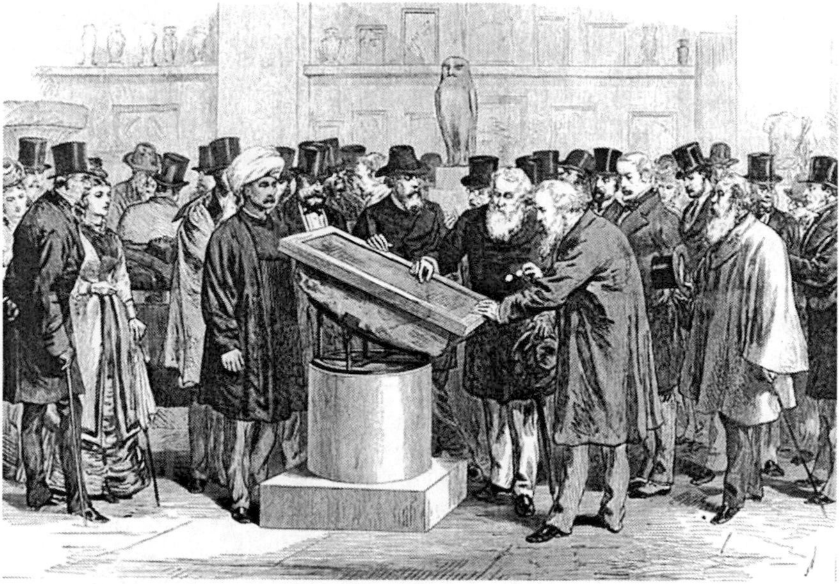

Figure 28.1 An illustration in the *London News* of Orientalists inspecting the Rosetta Stone during the International Congress of Orientalists, 1874

diverse women in distinct locations build transnational political projects.

Postmodern theorists also contributed to the concept of the other, but in a more celebratory than critical fashion. Postmodernism revelled in difference and in the deconstruction of binaries, essentially (and sometimes in essentializing fashion) celebrating 'otherness'. Postmodern perspectives on the city, for example, decentred the more powerful nodes around which urbanists had long understood urban processes, as in the centralized power of downtown financial districts. They celebrated, instead, a city with multiple nodes of power and belonging, a series of neighbourhoods where life and identity unfold in a more vertiginous and less orderly fashion (see Dear and Flusty 1997; Soja 1989). Rather than suggest a centre to any city surrounded by more peripheral and less important ('othered') neighbourhoods, the postmodern city is a geographical celebration of difference that moves sites once conceived of as 'marginal' to the centre of discussion and analysis.

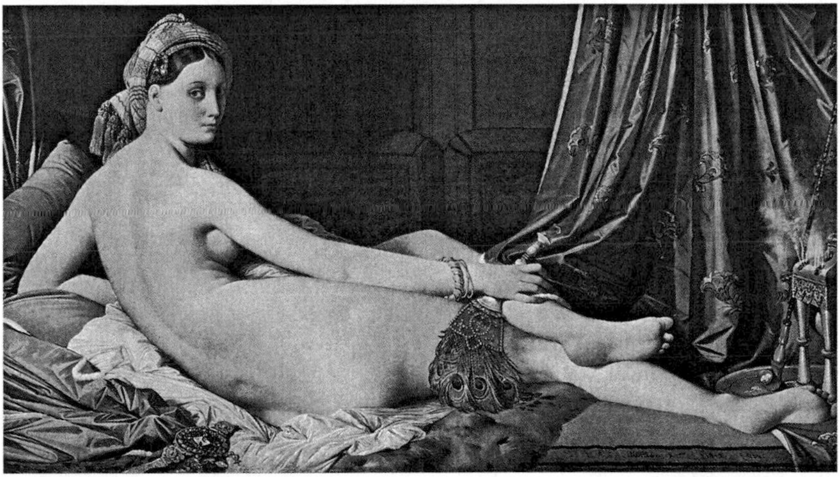

Figure 28.2 *The Grand Odalisque*, by Jean Auguste Dominique Ingres, 1814. Western artists often depicted Orientalist culture as sexually decadent. Although harems were quite rare in the 'Orient,' they were a particularly popular subject in Orientalist art.

Poststructural philosophies coming largely out of France by scholars including Jacques Derrida (1973), Jacques Lacan (1977), Michel Foucault (1970, 1991) and Julia Kristeva (1982) also produced insights on the 'other' in their theoretical examinations of difference. They challenged the idea of underlying truths, such as those set forth by Orientalists about 'the Orient'. They dwelt in particular on discourse analysis and the signification of language, seeing a key connection between discourses that 'other' people and the material effects that result. In particular, these scholars relied on an approach known as social constructionism, which critiques the idea that true meanings underlie language. Rather, this approach holds that to name (or categorize) something is to fix its meaning, often around a few essentialized attributes; to 'other' through the language of identification. A constructionist approach questions underlying meanings and referents and attempts to study and challenge the very categories that are used to 'other' groups of people, such as categories that racialize and categories central to Orientalism.

Critical race theorists, building on the work of poststructuralists, examined the construction of racialized categories that placed non-normative racial and ethnic markings and identities outside the norm. Like post-structuralists, they argued that 'race' did not exist as a biological and material phenomenon, but was a symbolic and linguistic act that essentialized and assigned attributes by category. Fierce debates ensued about whether 'race' was real, and how it mattered (e.g. West 1993). Under colonial regimes, for example, scientists for many years studied the physical characteristics of persons of different races for signs of differing traits. Measurements of differently sized crania were used by colonial powers, for example, to suggest differing levels of racial intelligence. In 1951, however, physical anthropologists working for the United Nations produced a report pronouncing that the differences between races were less significant than the commonalities or sameness across races, and hence all humans belonged to a single human race (Hyndman 2000). And yet, racialized categories continue as important measurements of difference. The United States government still takes a census of its population and asks individuals to distinguish themselves by race, thus reinforcing that race matters as a category of difference.

Immigrants and refugees are among those racialized and othered through categorization, differing legal status and public discourses such as news reports that characterize particular groups of immigrants. Often social scientists themselves do the othering by conducting studies in ways that homogenize and essentialize entire groups of people. By suggesting time and again that Mexicans are 'illegal', for example, public discourses advanced by media and politicians essentialize all Latin American immigrants as undocumented, and these representations promote discrimination against immigrants in the labour force and elsewhere.

Feminists, too, have a long tradition of examining the 'other'. While seeking the inclusion of women, feminists sought to simultaneously challenge the othering of women as non-normative figures. More recently, feminist scholars have studied identity more fully, a tradition that relies on concepts of 'self' and 'other' drawn from Freud, Lacan and Simone de Beauvoir, among others. Feminists departed from poststructuralists in their particular concern for equality between men and women and the inclusion of women in all facets of society. To provide an example, whereas a feminist project might examine the absence of women from politics, a poststructuralist project might examine how popular media represent women as inferior political leaders.

Cultural geographies of the 'other' rely heavily on spatial metaphors to locate and place identities and difference in the landscape. From

Gloria Anzaldúa's (1987) borderlands to bell hooks' (1991) margins, many cultural theorists write about locations with the use of spatial metaphors (Keith and Pile 1993). Likewise, political geographers have used the concept of the 'other' to examine how powerful countries dominate less powerful countries, regions or localities in an effort to legitimize their exploitation. In the US, for example, mainstream narratives present Appalachian people and culture as uncivilized, and the extraction of Appalachian resources such as coal as a civilizing process.

Queer theorists, meanwhile, celebrate 'othering' in that to 'queer' is to challenge normative categories. In this sense, queer theory does the work of displacing the centre in order to move 'other' from margin to centre. To queer the city, for example, would involve an examination of the sites where non-normative relationships take centre stage, whether in neighbourhoods or social settings where members of the lesbian, gay, bisexual and transgender (LGBT) community congregate to live, work, and challenge heteronormative expectations in so doing.

Case Study: Sexuality and Regionality

Case Study 1: Sexual Identity as 'Other'

Sexual and gender identities provide a dizzying array of standpoints from which to examine the concept of the 'other'. There are three (primary) intersecting axes along which gender and sexual identity are formed and simultaneously blurred. The first has to do with gender identity; the second with desire and attraction; the third with sex as it is biologically determined. Historically, people who move along these axes in their own unique and non-conventional ways have been considered sexual deviants. Women who love women, or men who love men, for example, were (and are still in particular times and places) deemed 'deviant' because of their attraction to persons of the same sex. They are 'others'; othered by a heteronormative world in which heterosexual couplings between men and women serve as the norm. Only over the last few decades has 'deviance' been changed in public discourse in order to normalize the reality that there are many ways of being in the world through sex, gender and desire. No one way is correct or normal.

Meanwhile, public policies designed by government, reinforced by faith-based organizations, and by social service and educational institutions, prioritize normative sexualities and sexual couplings. To provide one example, a person who suffers substance abuse and who identifies as a male-to-female transgender person may need to perform a gender identity with which she may not identify because treatment facilities generally segregate clients by sex: male or female, biologically determined.

While some individuals struggle in their journeys to seek social services, others struggle to acquire legal rights to maintain families viewed as 'non-traditional'. Minnie Bruce Pratt (2003: 52), poet, writer and activist, writes of losing custody of her children when she divorced her husband and entered into a relationship with a female partner. The assumption that a man and a woman in a heterosexual couple are the best parents for a child serves as an example of the power of heteronormativity in action. Minnie Bruce writes eloquently of the pain of othering, her long drives through the country to see her sons and the effects of heteronormative othering on family in her poem called 'Poem for my sons':

> That you'll remember me who crossed, recrossed
> you,
> as a woman making slowly toward an unknown place where you could be with me,
> like a woman on foot, in a long stepping out.

In spite of social movements that address the rights of members of the lesbian, gay, bisexual and transgender community, homosexuality is still considered a crime in many countries. Partners in same-sex couples face the tyranny of heteronormative public policies that govern families, marriage and adoption. Members of the LGBT community have embarked on social movements to fight for the legal right to marry, to adopt children, to be covered by a partner's health insurance plan and to have rights to make decisions about a partner's health care.

Geographers have researched how this othering manifests spatially in households and in migration patterns among same-sex couples (Cooke and Rapino 2007). David Bell and Gillian Valentine (1995) explore the experiences of men and women who find support and community around alterity and desire. Michael Brown (1997) studies how social movements arise and democracy functions in relation to the othering of persons with HIV/AIDS in Vancouver, British Columbia. The realities of othering influence where we live, how we live and how we survive.

Let's return to the example introduced earlier, the idea of 'queering' the city. Queer theorist Judith Halberstam (2005) turns to the city to

Figure 28.3 Gay pride parade in Toronto, Canada, 2005

look for examples of alternative or 'queer time' and 'queer space'. These are temporal and spatial patterns that challenge heteronormative assumptions and expectations about how life should work more generally. Cities are places where LGBT persons and identities manifest more visibly and freely in the landscape. In many cities, gay pride parades provide an outlet for 'queer time' (Figure 28.3). Another urban process that has been studied in association with queer identities is gentrification. Gentrification is the process wherein people invest in homes in urban neighbourhoods considered to be 'transitional'. Initially, individuals purchase and renovate homes. These individuals are often attracted to the diversity of the neighbourhood, a characteristic appealing and welcoming to 'alternate' (or othered) lifestyles (see Ley 1996; Smith 1996). As the neighbourhood improves, more commercial interests take root and capitalize on this alterity or difference, and the area becomes more expensive and more exclusive. The West Village in

Figure 28.4 The intersection of Hawley and Green Avenues in Syracuse, New York. The neighbourhood has been gentrified in part by members of the LGBT community. (Photo by and courtesy of Alison Mountz)

New York City, for example, in the 1960s was the centre of a generation of counter-cultural hippies and alternative lifestyles. In the decades since, it has become an exclusive, wealthy, expensive place to live. Those artists, students and hippies who originally made the neighbourhood 'hip' have long since been displaced. Yet the neighbourhood has maintained its reputation as a place where alternative or othered identities are celebrated, in part due to the commercial development and marketing of businesses and organizations that appeal to gay communities. In Syracuse, the Hawley-Green neighbourhood has also been gentrified by members of the LGBT community (Figure 28.4).

Case Study 2: Colonizing 'the Other'

Much of this discussion has addressed the discursive realm where 'the other' is conceptualized through knowledge production, but such processes are linked time and again to disastrous imperial projects in the international theatre. Not only are individuals 'othered' through practices surrounding gender and sexual identity, but entire populations are othered by performances of international relations, often through intersecting processes of racialization and sexualization. Far from abstract, or even benign, rationales associated with inferiority in relation to 'civilization' have brought states into armed conflict. The field of critical geopolitics (e.g. Dalby 1991; Ó Tuathail 1996a) examines the discursive production of 'others' and the ways that these knowledge productions inform international relations, in particular imperial and colonial projects.

In *The Colonial Present,* Derek Gregory (2004) argues that colonialism is alive and well. Gregory finds inspiration in Edward Said's orientalism and applies his ideas to contemporary rationales for the conflict known as 'the War on Terror' and used to justify the occupations of Afghanistan and Iraq. Drawing on the work of Said and Gregory, political geographer Karen Culcasi conducts a similar analysis of the ways that western colonial powers reproduced the region of the Middle East cartographically. Culcasi studied dozens of maps of the region produced by American and British cartographers alongside maps created by cartographers from the region, focusing especially on maps produced in and housed by Egypt. She finds that cartographers external to the region imposed an artificial regional identity on the landscape, one that illustrated 'flagrant disregard of the Arab right to self-determination' (Culcasi 2008: 265). Though artificial, the creation of 'the Middle East' as a region was then reproduced for decades to come. Culcasi finds that maps are but one part of a broader western discourse that creates the region as volatile, unpredictable, tumultuous and dangerous (Culcasi 2006; 2008: 1). Today, the Middle East continues to exist as a racialized location in the American and British geographical imagination.

In maps produced by cartographers internal to the region (and particularly in Egypt where she conducted research), Culcasi finds resistance to these characterizations. Maps by Arab cartographers illustrate the creation of an alternative, pan-Arab regional identity. Arab cartographers thus contested the colonial imaginings of the Middle East as 'other' with self-representation.

The Future of the 'Other'?

Whereas much research in the humanities and social sciences over the years has been premised on understanding 'the other' (Behar 1996), the hope is that those who have been conceptualized as 'other' will move to populate the centre; that the very concept of 'other' will eventually cease to exist. In this sense, 'other' is a verb. It acts. To other is to mark, separate, identify, discriminate, exclude or label a person or group as deviant. To other is to place power at the centre and deviance along the margins. With geographic vertigo, others and not others, centre and margin will swirl and blur, teeter and twirl. Studies of race and sexuality, gender and class, ability and disability will not be cast off to the margins of university disciplines in the social sciences, but will infuse and inform them.

KEY POINTS

- The term 'other' serves as both a noun and a verb. As a noun, the 'other' is a person or group of people different from oneself. As a verb, 'other' means to distinguish, label, categorize, name, identify, place and exclude those who do not fit a societal norm.
- In geographic terms to 'other' is to locate a person or group of persons outside the centre, on the margins.
- Geographers working in the postcolonial tradition were the first to widely use the concept of the 'other'. Political geographers used the concept to explain how colonial and imperial powers 'othered' places they wanted to dominate in order to legitimize their exploitation.
- Social movements led by 'othered' groups use space to challenge normative assumptions about the way life should work. Gay pride parades, for example, challenge the idea that gay people should hide their sexual orientation.

FURTHER READING

Brown, M. (2000) *Closet Spaces: Geographies of Metaphor from the Body to the Globe*. New York: Routledge.

Gregory, D. (2004) *The Colonial Present: Afghanistan, Palestine, Iraq*. Malden, MA: Blackwell.

Gregory, D. and Pred, A. (eds) (2006) *Violent Geographies: Fear, Terror and Political Violence*. New York: Routledge.

Mohanty, C. (2003) *Feminism without Borders: Decolonizing Theory, Practicing Solidarity*. Durham, NC: Duke University Press.

Said, E. (1978) *Orientalism*. New York: Pantheon Books.

BIBLIOGRAPHY

Abu-Loghod, J. (1989) *Before European Hegemony: The World System AD 1250–1350*. New York: Oxford University Press.

Achcar, G. (2000) 'The strategic triad: USA, China, Russia', in T. Ali (ed.), *Masters of the Universe? NATO's Balkan Crusade*. New York: Verso. pp. 99–144.

Adamson, W. (1980) 'Gramsci's interpretation of fascism', *Journal of the History of Ideas*, 41(4): 615–633.

Agamben, G. (1998) *Homo Sacer: Sovereign Power and Bare Life*. Stanford: Stanford University Press.

Aglietta, M. (1979) *A Theory of Capitalist Regulation: The US Experience*. Translated by D. Fernback. London: New Left Books.

Agnew, J. (1997) 'The dramaturgy of horizons: geographical scale in the reconstruction of Italy by the new Italian political parties, 1992–95', *Political Geography*, 16(2): 99–121.

Agnew, J. (1998) *Geopolitics: Re-visioning World Politics*. New York: Routledge.

Agnew, J. (2001) 'Regions in revolt', *Progress in Human Geography*, 25(1): 103–110.

Agnew, J. (2002) *Making Political Geography*. London: Arnold.

Agnew, J. (2003a) 'Contemporary political geography: intellectual heterodoxy and its dilemmas', *Political Geography*, 22: 603–606.

Agnew, J. (2003b) 'American hegemony into American empire? Lessons from the invasion of Iraq', *Antipode* 35(5): 871–885.

Agnew, J. (2003c) *Geopolitics: Re-visioning world politics*, 2nd edn. New York: Routledge.

Agnew, J. (2005) 'Sovereignty regimes: territoriality and state authority in contemporary world politics', *Annals of the Association of American Geographers*, 95(2): 437–461.

Agnew, J. (2006) 'Religion and geopolitics', *Geopolitics*, 11(2): 183–191.

Agnew, J. and Corbridge, S. (1995) *Mastering Space: Hegemony, Territory and International political Economy*. London: Routledge.

Agnew, J., Mitchell, K. and Toal, G. (eds) (2003a) *A Companion to Political Geography*. Malden, MA: Blackwell.

Agnew, J., Mitchell, K. and Toal, G. (2003b) 'Introduction', in J. Agnew, K. Mitchell and G. Toal (eds), *A Companion to Political Geography*. Malden, MA: Blackwell. pp. 1–19.

Aitchison, C., Hopkins, P. and Kwan, M. (eds) (2007) *Geographies of Muslim Identities: Diaspora, Gender and Belonging*. Aldershot: Ashgate.

Alcoff, L.M. (2006) *Visible identities: Race, Gender, and the Self*. Oxford: Oxford University Press.

Amin, A. and Thrift, N. (2005) 'What's left? Just the future', *Antipode*, 37(2): 220–238.

Amin, A. and Thrift, N. (2007) 'On being political', *Transactions of the Institute of British Geographers*, 32(1): 112–115.

Amnesty International (2003) '"Our brothers who help kill us" – economic exploitation and human rights abuses in the east'. Retrieved 31 July 2007, from:

http://www.amnestyusa.org/document.php?lang=e&id=A4B3F753111D01D28025 6D19004492FF

Amoore, L. (2006) 'Biometric borders: governing mobilities in the war on terror', *Political Geography*, 25(3): 336–351.

Anderson, B. (1983) *Imagined Communities: Reflections on the Origins and Spread of Nationalism*. London and New York: Verso.

Anderson, J. L. (2007) 'The Taliban's opium war: the difficulties and dangers of the eradication program', *The New Yorker*, 83(19), 9 July: 60–71.

Andreas, P. (2000) *Border Games: Policing the US–Mexico Divide*. Ithaca, NY: Cornell University Press.

Anzaldúa, G. (1987) *Borderlands/La Frontera: The New Mestiza*. San Francisco, CA: Aunt Lute.

Appadurai, A. (1996) *Modernity at Large: Cultural Dimensions of Globalization*. Minneapolis: University of Minnesota Press.

Arendt, H. (1958) *The Origins of Totalitarianism*. Cleveland and New York: Meridian Books.

Arvidson, E. (2000) 'Los Angeles: a postmodern class mapping', in J.K. Gibson-Graham, S. Resnick and R. Wolff (eds), *Class and its Others*. Minneapolis: University of Minnesota Press. pp. 163–190.

Assies, W. (2003) 'David versus Goliath in Cochabamba: water rights, neoliberalism, and the revival of social protest in Bolivia', *Latin American Perspectives*, 30(3): 14–36.

Axell, A. and Kase, H. (2002) *Kamikazi: Japan's Suicide Gods*. New York: Longman.

Bailey, A. (2005) *Making Population Geography*. London: Hodder Arnold.

Bailey, A., Wright, R., Mountz, A. and Miyares, I. (2002) '(Re)producing Salvadoran transnational geographies', *Annals of the Association of American Geographers*, 92(1): 125–144.

Baram, A. (1997) 'Neo-tribalism in Iraq: Saddam Hussein's tribal policies 1991–96', *International Journal of Middle East Studies*, 29(1): 1–31.

Barash, D. (2000) *Approaches to Peace: A Reader in Peace Studies*. New York: Oxford University Press.

Barber, B. (1996) *Jihad vs. McWorld: How Globalism and Tribalism are Reshaping the World*. New York: Ballantine Books.

Barnett, C. (1997) 'Sing along with the common people: politics, postcolonialism and other figures', *Environment and Planning D: Society and Space*, 15: 137–154.

Barnett, C. and Low, M. (eds) (2004) *Spaces of Democracy*. London: Sage.

Basch, L., Glick-Schiller, N. and Blanc-Szanton, C. (1994) *Nations Unbound: Transnational Projects, Postcolonial Predicaments, and Deterritorialized Nation-states*. Langhorne: Gordon and Breach.

Bassin, M. (1987) 'Race contra space: the conflict between German geopolitik and National Socialism', *Political Geography Quarterly*, 6: 115–134.

Bechtel Corporation (2006) *Cochabamba Water Dispute Settled*. Bechtel Corporation press release, 19 January. Retrieved 11 October 2007, from: http://www.bechtel.com/newsarticles/487.asp

Beck, R. (2003) 'Remote sensing and GIS as counterterrorism tools in the Afghanistan War: a case study of the Zhawar Kili region', *The Professional Geographer*, 55(2): 170–179.

Behar, R. (1996) *The Vulnerable Observer*. Boston, MA: Beacon Press.

Bell, D. and Valentine, G. (eds) (1995) *Mapping Desire: Geographies of Sexualities*. London and New York: Routledge.

Bell, J.E. and Staeheli, L. (2001) 'Discourses of diffusion and democratization', *Political Geography*, 20: 175–195.

Bennett, R. (1998) 'Business associations and their potential to contribute to economic development: re-exploring an interface between state and market', *Environment and Planning A*, 30: 1367–1387.

Berg, L., Evans, M., Fuller, D. and The Okanagan Urban Aboriginal Health Research Collective Canada (2007) 'Ethics, hegemonic whiteness, and the contested imagination of "aboriginal community" in social science research in Canada', *ACME: An International E-Journal for Critical Geographies*, 6(3): 395–410. Retrieved 5 January 2008, from: http://www.acme-journal.org/vol6/LDBetal.pdf

Bhabha, H. (1990) *Nation and Narration*. New York and London: Routledge.

Bialasiewicz, L., Campbell, D., Elden, S., Graham, S., Jeffrey, A. and Williams, A.J. (2007) 'Performing security: the imaginative geographies of current US strategy', *Political Geography*, 26(4): 405–422.

Bigo, D. (2002) 'Security and immigration: toward a critique of the governmentality of unease', *Alternatives*, 27: 63–92.

Binnendijk, H. and Kugler, R. L. (2001) *Revising the Two-Major Theater War Standard*. Washington, DC: Institute for National Strategic Studies.

Blaikie, P. and Brookfield, H. (eds) (1987) *Land Degradation and Society*. London: Methuen.

Blaut, J. M. (1993) *The Colonizer's Model of the World: Geographical Diffusionism and Eurocentric History*. New York and London: Guilford Press.

Blaut, J. M. (1999) 'Environmentalism and eurocentrism', *Geographical Review*, 89(3): 391–408.

Blaut, J. M. (2000) *Eight Eurocentric Historians*. New York: Guilford Press.

Blunt, A. (1994) *Travel, Gender and Imperialism: Mary Kingsley and West Africa*. New York and London: Guilford Press.

Blunt, A. and McEwan, C. (eds) (2002) *Postcolonial Geographies*. London and New York: Continuum.

Blunt, A. and Rose, G. (eds) (1994) *Writing Women and Space: Colonial and Postcolonial Geographies*. New York: Guilford Press.

Blunt, A. and Wills, J. (2000) *Dissident Geographies: An Introduction to Radical Ideas and Practice*. Harlow: Pearson Education.

Boal, F. (1996) 'Integration and division: sharing and segregating in Belfast', *Planning Practice and Research*, 11(2): 151–158.

Bondi, L. (1990) 'Feminism, postmodernism, and geography: space for women?' *Antipode*, 22: 156–167.

Bondi, L. (1993) 'Locating identity politics', in M. Keith and S. Pile (eds), *Place and the Politics of Identity*. New York: Routledge. pp. 84–101.

Borja, J. and Castells, M. (1997) *The Local and the Global: The Management of Cities in the Information Age* . London: James and James/Earthscan.

Bowman, I. (1921) *The New World: Problems in Political Geography*. Yonkers on the Hudson: World Book Company.

Boyle, M. and Rogerson, R. (2006) '"Third Way" Urban Policy and the New Moral Politics of Community', *Urban Geography*, 27(3): 201–227.

Braden, K. and Shelley, F. (2000) *Engaging Geopolitics*. Harlow: Pearson Education.

Braudel, F. (1993) *A History of Civilizations*, translated from the French by Richard Maynes. New York: Penguin Classics.

Breitbart, M. (1975) 'Impressions of an anarchist landscape', *Antipode*, 7: 44–49.

Brenner, N. (1998) 'Between fixity and motion: accumulation, territorial organization and the historical geography of spatial scales', *Environment and Planning D: Society and Space*, 16: 459–481.

Bridge, G. (2002) 'The grounding globalization: the prospects and perils of linking economic processes of globalization to environmental outcomes', *Economic Geography*, 78(3): 361–386.

Brinegar, J. (2001) 'Female representation in the discipline of geography', *Journal of Geography in Higher Education*, 25: 311–320.

Brown, M. (1997) *RePlacing Citizenship: AIDS Activism and Radical Democracy*. New York: Guilford Press.

Brown, M. (2000) *Closet Spaces: Geographies of Metaphor from the Body to the Globe*. New York: Routledge.

Brown, M., Knopp, L. and Morrill, R. (2004) 'The culture wars and urban electoral politics: sexuality, race, and class in Tacoma, Washington', *Political Geography*, 24(3): 267–291.

Brownlie, I. (1998) *Principles of Public International Law*, 5th edn. New York: Oxford University Press.

Bunge, W. (1969) *Atlas of Love and Hate*. Detroit: The Society for Human Exploration.

Butler, J. (2004) *Undoing Gender*. London and New York: Routledge.

Canetti-Nisim, D., Mesch, G. and Pedahzur, A. (2006) 'Victimization from terrorist attacks: randomness or routine activities?' *Terrorism and Political Violence*, 18: 485–501.

Caplan, R. (2002) *A New Trusteeship? The International Administration of War-torn Territories*. New York: Oxford University Press.

Castells M. (1996) *The Rise of the Network Society*. Oxford: Blackwell.

Castells, M. (1997) *The Power of Identity*. Malden, MA: Blackwell.

Castells, M. (1998) *The Power of Identity: The Information Age: Economy, Society, and Culture, Volume II*. New York: Blackwell.

Castells, M. (2001) 'Information technology and global capitalism', in W. Hutton and A. Giddens (eds), *On the Edge: Living with Global Capitalism*. London: Vintage.

Castles, S. and Miller, M. (1998) *The Age of Migration: International Population Movements in the Modern World*. New York and London: Guilford Press.

Chin, K. L. (1999) *Smuggled Chinese: Clandestine Immigration to the United States*. Philadelphia: Temple University Press.

Chossudovsky, M. (1997) *The Globalization of Poverty: Impacts of the IMF and World Bank Reforms*. London: Zed Books.

Chouinard, V. and Crooks, V. (2005) '"Because *they* have all the power and I have none": state restructuring of income and employment supports and disabled women's lives in Ontario, Canada', *Disability and Society*, 20(1): 19–32.

Chua, A. (2007) *Day of Empire: How Hyperpowers Rise to Global Dominance – and Why they Fall.* New York: Doubleday.

Clayton, D. (2000) *Island of Truth: The Imperial Fashioning of Vancouver Island.* Vancouver: University of British Columbia Press.

Cleary, M. (2006) 'A "left turn" in Latin America? Explaining the left's resurgence', *Journal of Democracy*, 17(4): 35–49.

Clinton, W. J. (1996) State of the Union Address, 23 January. Retrieved 15 August 2008 from the website of the William J. Clinton Foundation: http://www.clinton foundation.org/video.htm?title=State%20of%20the%20Union,%201996

Coakley, J. and O'Dowd, L. (eds) (2007) 'Partition and the reconfiguration of the Irish border', *Political Geography*, 26(8): 877–982.

Coleman, M. (2007) 'Immigration geopolitics beyond the US–Mexico border', *Geopolitics*, 39(1): 54–76.

Cooke, T. and Rapino, M. (2007) 'The migration of partnered gays and lesbians between 1995 and 2000', *Professional Geographer*, 59(3): 285–297.

Cope, M. (2004) 'Placing gendered political acts', in L. Staeheli, E. Kofman and L. Peake (eds), *Mapping Women, Making Politics: Feminist Perspectives on Political Geography*. London and New York: Routledge. pp. 71–86.

Cowen, D. and Gilbert, E. (2007) 'Citizenship in the "homeland": families at war?' in D. Cowen and E. Gilbert (eds), *War, Citizenship, Territory*. New York: Routledge.

Cox, K. (2002) *Political Geography: Territory, State and Society*. Oxford: Blackwell.

Cox, K. (ed.) (2003) 'Forum: political geography in question', *Political Geography*, 22(6): 599–675.

Cox, K., Low, M. and Robinson, J. (eds) (2007) *A Handbook of Political Geography*. Thousand Oaks, CA: Sage.

Crampton, A. (2001) 'The Voortrekker Monument, the birth of apartheid, and beyond', *Political Geography*, 20(2): 221–246.

Crenshaw, M. (1981) 'The causes of terrorism', *Comparative Politics*, 13(4): 379–399.

Cresswell, T. (1996) *In Place/Out of Place: Geography, Ideology and Transgression.* Minneapolis: University of Minnesota Press.

Culcasi, Karen (2006) 'Cartographically constructing Kurdistan within geopolitical and orientalist discourses', *Political Geography*, 25(6): 680–706.

Culcasi, Karen (2008) Unpublished PhD Dissertation. Syracuse University, New York.

Cutter, S., Richardson, D. and Wilbanks, T. (eds) (2003) *The Geographical Dimensions of Terrorism*. New York: Routledge.

Dahlman, C. T. (2002) 'The political geography of Kurdistan', *Eurasian Geography and Economics*, 43(4): 271–299.

Dahlman, C. T. (2004) 'Turkey's accession to the European Union: the geopolitics of enlargement', *Eurasian Geography and Economics*, 45(8): 553–574.

Dahlman, C. T. and Ó Tuathail, G. (2005) 'The legacy of ethnic cleansing: the international community and the returns process in post-Dayton Bosnia-Herzegovina', *Political Geography*, 24(5): 569–599.

Dalby, S. (1991) 'Critical geopolitics: discourse, difference, and dissent', *Environment and Planning D: Society and Space*, 9(3): 261–283.

Dalby, S. (1998) *Creating the Second Cold War: The Discourse of Politics*. New York: Guilford Press.

Dalby, S. (2002) *Environmental Security*. Minneapolis: University of Minnesota Press.

Dear, M. and Flusty, S. (1997) 'The iron lotus: Los Angeles and postmodern urbanism', *Annals of the American Academy of Political and Social Science*, 551(1): 151–163.

de Blij, H.J. (2004) 'Explicating geography's dimensions – an opportunity missed', *Annals of the Association of American Geographers,* 94(4): 994–996.

de Waal, T. (2003) *Black Garden: Armenia and Azerbaijan through Peace and War.* New York: New York University Press.

Delaney, D. and Leitner, H. (1997) 'The political construction of scale', *Political Geography*, 16(2): 93–97.

Derrida, J. (1973) *Speech and Phenomena and Other Essays on Husserl's Theory of Signs*. Evanston: Northwestern University Press.

Desbiens, C., Mountz, A. and Walton-Roberts, M. (2004) 'Re-thinking the state from the margins of political geography', *Political Geography*, 23(3): 241–243.

Des Forges, R. and Xu, L. (2001) 'China as a non-hegemonic superpower? The uses of history among the China can say no writers and their critics', *Critical Asian Studies*, 33(4): 483–507.

Diamond, J. (1997) *Guns, Germs, and Steel: The Fates of Human Societies.* New York: W.W. Norton.

Dicken, P. (2003) *Global Shift.* New York and London: Guilford Press.

Dodds, K. (2000) *Geopolitics in a Changing World*. New York: Prentice-Hall.

Dodds, K. (2005) *Global Geopolitics: A Critical Introduction.* Harlow: Pearson Education.

Dodge, T. (2003) *Inventing Iraq: The Failure of Nation Building and a History Denied.* New York: Columbia University Press.

Domosh, M. and Seager, J. (2001) *Putting Women in Place: Feminist Geographers Make Sense of the World.* New York and London: Guilford Press.

Dooley, P. (2005) *The Labor Theory of Value*. New York: Routledge.

Dower, J. (2000) *Embracing Defeat: Japan in the Aftermath of World War II.* New York: Penguin Press.

Dowler, L. and Sharp, J. (2001) 'A feminist geopolitics?' *Space and Polity*, 5(3): 165–176.

Dowling, R. and Blunt, A. (2006) *Home.* London: Routledge.

Driver, F. (1999) *Geography Militant: Cultures of Exploration in the Age of Empire.* Oxford: Blackwell.

Dwyer, C. (2002) '"Where are you from?" Young British Muslim women and the making of "home"', in A. Blunt and C. McEwan (eds), *Postcolonial Geographies.* New York: Guilford Press. pp. 184–199.

Dyck, I., Lewis, N. D. and McLafferty, S. (eds) (2001) *Geographies of Women's Health.* London and New York: Routledge.

Dyer, J. (1998) *Harvest of Rage: Why Oklahoma City is Only the Beginning*. Boulder, CO: Westview.

Dyke, C. (1988) *The Evolutionary Dynamics of Complex Systems*. New York: Oxford University Press.

Eagleton, T. (1991) *Ideology: An Introduction*. London and New York: Verso.

Eck, K., Sollenberg, M. and Wallensteen, P. (2004) 'One-sided violence and non-state conflict', in L. Harbom (ed.), *States in Armed Conflict.* Uppsala: Department of Peace and Conflict Research. pp. 133–142.

Eley, G. (2002) *Forging Democracy: The History of the Left in Europe 1850–2000*. New York: Oxford University Press.

England, K. (2003) 'Towards a feminist political geography?' *Political Geography*, 22(6): 611–616.

Ettlinger, N. (2007) 'Bringing democracy home: post-Katrina New Orleans', *Antipode*, 39(1): 8–16.

Ettlinger, N. and Bosco, F. (2004) 'Thinking through networks and their spatiality: a critique of the US (public) war on terrorism and its geographic discourse', *Antipode*, 36(2): 249–271.

Fabrizio, E. (2007) '"Reclus versus Ratzel": from state geopolitics to human geopolitics', *Research in Anarchism Forum*, Article 205, 2 December. Retrieved 26 July 2007, from http://raforum.info/article.php3?id_article=3040

Falah, G-W. (ed.) (2003) Forum on the 2003 War on/in Iraq. *Arab World Geographer*, 6(1). Retrieved 1 February 2008, from http://users.fmg.uva.nl/vmamadouh/awg/

Falah, G-W. (ed.) (2005) Forum on the viability of a two-state solution to the Israel/Palestine conflict and possible alternatives? *Arab World Geographer*, 8(3). Retrieved 1 November 2007 from http://users.fmg.uva.nl/vmamadouh/awg/

Falah, G-W. and Flint, C. (2004) Geopolitical spaces: the dialectic of public and private space in the Palestine–Israel conflict. *Arab World Geographer*, 7(1–2). Retrieved 1 November 2007, from http://users.fmg.uva.nl/vmamadouh/awg/

Falah, G-W. and Nagel, C. (eds) (2005) *Geographies of Muslim Women: Gender, Religion, and Space*. New York: Guilford Press.

Fan, C. (2004) 'The state, the migrant labor regime, and maiden workers in China', *Political Geography*, 23: 283–305.

Featherstone, M. (1990) *Global Culture: Nationalism, Globalization, and Modernity*. London: Sage.

Feinberg, L. (1998) *Trans Liberation: Beyond Pink or Blue*. Boston, MA: Beacon Press.

Feitelson, E. and Levy, N. (2006) 'The environmental aspects of reterritorialization: environmental facets of Israeli–Arab agreements', *Political Geography*, 25(4): 459–477.

Feldman, G. (2005) 'Essential crises: a performative approach to migrants, minorities, and the European nation-state', *Anthropological Quarterly*, 78(1): 213–246.

Fiori, G. (1996) *Antonio Gramsci: Life of a Revolutionary*. London: Verso Modern Classics.

Fishman, T. C. (2006) *China, Inc.: How the Rise of the Next Superpower Challenges America and the World*. New York: Scribner.

Flint, C. (2001) 'Right-wing resistance to the process of American hegemony: the changing political geography of nativism in Pennsylvania, 1920–1998', *Political Geography*, 20(6): 763–786.

Flint, C. (2002) 'Political geography: globalization, metapolitical geographies and everyday life', *Progress in Human Geography*, 26(3): 391–400.

Flint, C. (2003a) 'Dying for a "P"? Some questions facing contemporary political geography', *Political Geography*, 22: 617–620.

Flint, C. (2003b) 'Terrorism and counterterrorism: geographic research questions and agendas', *The Professional Geographer*, 55(2): 161–169.

Flint, C. (ed.) (2005) *The Geography of War and Peace: From Death Camps to Diplomats*. Oxford: Oxford University Press.

Flint, C. and Shelley, F. (1990) 'Structure, agency, and context: the contributions of geography to world-systems analysis', *Sociological Inquiry*, 60(1): 496–508.

Foot, R. (2006) 'Chinese strategies in a US-hegemonic global order: accommodating and hedging', *International Affairs*, 82(1): 77–94.

Foucault, M. (1970) *The Order of Things: An Archaeology of the Human Sciences*. London: Tavistock.

Foucault, M. (1977) *Discipline and Punish: The Birth of the Prison*. New York: Vintage Books.

Foucault, M. (1991) *The History of Sexuality, Volume 1*. New York: Vintage.

Foucault, M. (2003) *'Society must be defended': Lectures at the Collège de France 1975–1976*. New York: Picador.

Fox, W. T. R. (1944) *The Super-powers: The United States, Britain, and the Soviet Union – Responsibility for Peace*. New York: Harcourt Brace.

Frank, A. G. (1969) *Capitalism and Underdevelopment in Latin America : Historical Studies of Chile and Brazil*. New York: Monthly Review Press.

Fraser, A. (2008) 'White farmers' dealings with land reform in South Africa: evidence from northern Limpopo province', *Tijdschrift voor Economishe en Sociale Geografie*, 99(1): 24–36.

Freeman, C. (2001) 'Is local:global as feminine:masculine? Rethinking the gender of globalization', *Signs*, 26(4): 1007–1037.

Friedman, M. (1962) *Capitalism and Freedom*. Chicago: University of Chicago Press.

Fröebel, F., Heinricks, J. and Kreye, O. (1980) *The New International Division of Labor*. Cambridge: Cambridge University Press.

Froehling, O. (1999) 'Internauts and guerilleros: the Zapatista rebellion in Chiapas, Mexico and its extension into cyberspace', in P. Crang and J. May (eds), *Virtual Geographies: Bodies, Space and Relations*. New York: Routledge. pp. 164–177.

Frontline (2002) *Bolivia: Leasing the Rain. Timeline Cochabamba Water Revolt*. Retrieved 12 October 2007, from http://www.pbs.org/frontlineworld/stories/bolivia/timeline.html

Frum, D. and Perle, R. (2004) *An End to Evil: How to Win the War on Terror*. New York: Ballantine Books.

Fukuyama, F. (1989) 'The end of history', *National Interest*, Summer.

Fukuyama, F. (2004) *State-building: Governance and World Order in the 21st Century*. Ithaca, NY: Cornell University Press.

Gaddis, J. L. (2005) *The Cold War*. New York: Penguin Press.

Gagnon, V. P. (2006) *The Myth of Ethnic War: Serbia and Croatia in the 1990s*. Ithaca, NY: Cornell University Press.

Gallaher, C. (2003) *On the Fault Line: Race, Class, and the American Patriot Movement*. Lanham, MD: Rowman and Littlefield.

Gallaher, C. (2004) 'Teaching about political violence: a primer on representation', *Journal of Geography in Higher Education*, 28(2): 301–315.

Gallaher, C. and Froehling, O. (2002) 'New world warriors: "nation" and "state" in the politics of the zapatista and US patriot movements', *Social and Cultural Geography*, 3(1): 81–102.

Galloway, G. (2003) 'Emergency preparedness and response – lessons learned from 9/11', in S. Cutter, D. Richardson and T. Wilbanks (eds), *The Geographical Dimensions of Terrorism*. New York: Routledge. pp. 27–34.

Galtung, J. (1990) 'Cultural violence', *Journal of Peace Research*, 27(3): 291–305.

Geopolitics (2007) Oxford: Oxford University Press.

George, S. (1990) *A Fate Worse than Debt: The World Financial Crisis and the Poor*. New York: Grove Weidenfeld.

Gibson-Graham, J.K. (1996) *The End of Capitalism (As We Knew It): A Feminist Critique of Political Economy*. Minneapolis: University of Minnesota Press.

Gibson-Graham, J.K., Resnick, S. and Wolff, R. (2000) 'Class in a poststructuralist frame', in J.K. Gibson-Graham, S. Resnick, and R. Wolff (eds), *Class and Its Others*. Minneapolis: University of Minnesota Press. pp. 1–22.

Giddens, A. (1990) *The Consequences of Modernity*. Cambridge: Polity Press.

Giddens, A. (2006) *Sociology* (5th edn). Cambridge: Polity.

Giles, W. and Hyndman, J. (eds) (2004) *Sites of Violence: Gender and Conflict Zones*. Berkeley: University of California Press.

Gilmartin, M. and Berg, L. (2007) 'Locating postcolonialism', *Area*, 39(1), 120–124.

Giraut, F. and Maharaj, B. (2002) 'Contested terrains: cities and hinterlands in post-apartheid boundary delimitations', *Geojournal*, 57(1–2): 39–51.

Glick-Schiller, N., Basch, L. and Blanc-Szanton, C. (eds) (1992) *Towards a Transnational Perspective on Migration: Race, Class, Ethnicity, and Nationalism Reconsidered*. New York: New York Academy of Sciences.

Godlewska, A. and Smith, N. (eds) (1994) *Geography and Empire*. Malden, MA: Blackwell.

Golledge, R. G., Church, R., Dozier, J., Estes, J. E., Michaelsen, J., Simonett, D. S., Smith, R., Smith, T., Strahler, A. H. and Tobler, W. R. (1982) Commentary on 'The Highest form of the Geographer's, *Annals of the Association of American Geographers* Art 72(4): 557–558.

Gomez-Peña, G. (1996) *The New World Border: Prophecies, Poems and Loqueras for the End of the Century*. San Francisco, CA: City Lights Books.

Gooder, H. and Jacobs, J. (2002) 'Belonging and non-belonging: the apology in a reconciling nation', in A. Blunt and C. McEwan (eds), *Postcolonial Geographies*. New York: Guilford Press. pp. 200–213.

Goss, J. (1995) 'Marketing the new marketing: the strategic discourse of geodemographic information systems', in J. Pickles (ed.), *Ground Truth: The Social Implications of Geographic Information Systems*. New York: Guilford Press. pp. 130–170.

Gottesman, E. R. (2004) *Cambodia after the Khmer Rouge: Inside the Politics of Nation Building*. New Haven, CT: Yale University Press.

Gottmann, J. (1973) *The Significance of Territory*. Charlottesville: University of Virginia Press.

Graham, B. and Shirlow, P. (2002) 'The Battle of the Somme in Ulster memory and identity', *Political Geography*, 21: 881–904.

Graham, E. (1999) 'Breaking out: the opportunities and challenges of multi-method research in population geography', *Professional Geographer*, 51(1): 76–89.

Gramsci, A. (1971) *Selections from the Prison Notebooks of Antonio Gramsci*, edited and translated from the Italian by Quintin Hoare and Geoffrey Smith. New York: International Publishers.

Gregory, D. (1978) *Ideology, Science and Human Geography*. London: Hutchinson.

Gregory, D. (ed.) (2000) *The Dictionary of Human Geography*. Malden, MA: Blackwell.

Gregory, D. (2004) *The Colonial Present: Afghanistan, Palestine, Iraq*. Malden, MA: Blackwell.

Gregory, D. and Pred, A. (eds) (2006) *Violent Geographies: Fear, Terror and Political Violence*. New York: Routledge.

Grewal, I. (2003) 'Transnational America: race, gender and citizenship after 9/11', *Social Identities*, 9(4): 535–561.

Griffith, D. (2004) Using Maps to Plug Security Gaps: Fact or Fantasy? *Annals of the Association of American Geographers*, 94(4): 998–1001.

Guevara, E. (1998) *Guerilla Welfare*, Marc Becker Contributor. Lincoln: University of Nebraska Press.

Haj, S. (1997) *The Making of Iraq, 1900–1963: Capital, Power and Ideology*. Albany: State University of New York Press.

Halal, W. (1986) *The New Capitalism*. New York: John Wiley and Sons.

Halberstam, J. (2005) *In a Queer Time & Place: Transgender Bodies, Subcultural Lives*. New York: New York University Press.

Hall, S. (1996) When was the 'postcolonial'? Thinking at the limit. In I. Chambers and L. Curti (eds), *The Post-Colonial Question: Common Skies, Divided Horizons*. London and New York: Routledge, pp. 242–260.

Hannah, M. (2000) *Governmentality and the Mastery of Territory in Nineteenth-century America*. Cambridge: Cambridge University Press.

Hanson, S. and Pratt, G. (1995) *Gender, Work and Space*. New York and London: Routledge.

Harding, A. (2000) *Is There a 'Missing Middle' in English Governance?* London: New Local Government Network.

Hardt, M. and Negri, A. (2004) *Multitude: War and Democracy in the Age of Empire*. New York: Penguin Books.

Harris, C. (2002) *Making Native Space*. Vancouver: University of British Columbia Press.

Hart, G. (2002) *Disabling Globalisation: Places of Power in Post-apartheid South Africa*. Berkeley: University of California Press.

Hart, J.F. (1982) The Highest form of the Geographer's Art. *Annals of the Association of American Geographers*, 72(1): 1–29.

Hartshorne, R. (1939) 'The nature of geography', *Annals of the Association of American Geographers*, 29: 1–28.

Hartshorne, R. (1950) 'The functional approach in political geography', *Annals of the Association of American Geographers*, 40(2): 95–130.

Hartshorne, R. (1954) 'Comment on "exceptionalism in geography"', *Annals of the Association of American Geographers*, 44: 108–109.

Hartshorne, R. (1955) '"Exceptionalism in geography" reexamined', *Annals of the Association of American Geographers*, 45: 205–244.

Harvey, D. (1973) *Social Justice and the City*. Baltimore, MD: Johns Hopkins University Press.

Harvey, D. (1982) *The Limits to Capital*. Chicago: University of Chicago Press.

Harvey, D. (1989) *The Condition of Postmodernity*. Cambridge, MA: Blackwell.

Harvey, D. (2000) *Spaces of Hope*. Berkeley: University of California Press.

Harvey, D. (2001) *Spaces of Capital*. New York: Routledge.

Harvey, D. (2005) *The New Imperialism*, 2nd edn. New York: Oxford University Press.

Harvey, D. (2003) *The New Imperialism*. Oxford: Oxford University Press.

Harvey, D. (2006) 'The geographies of critical geography', *Transactions of the Institute of British Geographers*, 31(4): 409–412.

Harvey, D. (2007) *A Brief History of Neoliberalism*. Oxford: Oxford University Press.

Healey, R. (1983) Regional Geography in the Computer Age: A Further Commentary on 'The Highest form of the Geographer's Art'. *Annals of the Association of American Geographers*, 73(3): 439–441.

Held, D., McGrew, A., Goldblatt, D. and Perration, J. (1999) *Global Transformations: Politics, Economics, and Culture*. Stanford, CA: Stanford University Press.

Herb, G. H. and Kaplan, D. H. (1999) *Nested Identities: Nationalism, Territory, and Scale*. Lanham, MD: Rowman and Littlefield.

Herbert, S. (2007) 'The "battle of Seattle" revisited: or, seven views of a protest-zoning state', *Political Geography*, 26(5): 601–619.

Herod, A. (2000) 'Implications of just-in-time production for union strategy: lessons from the 1998 General Motors–United Auto Workers dispute', *Annals of the Association of American Geographers*, 90: 521–547.

Herrschel, T. (2007) 'Between difference and adjustment – the re-/presentation and implementation of post-socialist (communist) transformation', *Geoforum*, 38(3): 429–444.

Herz, J. (1957) 'Rise and demise of the territorial state', *World Politics*, 9(4): 473–493.

Hobsbawm, E. (1990) *Nations and Nationalism since 1780*. Cambridge: Cambridge University Press.

Hoffman, B. (1998) *Inside Terrorism*. London: Indigo.

Holzgrefe, J. L. (2003) 'The humanitarian intervention debate', in J. L. Holzgrefe and R. O. Keohane (eds), *Humanitarian Intervention: Ethical, Legal and Political Dilemmas*. Cambridge: Cambridge University Press. pp. 15–52.

Homer-Dixon, T. (2001) *Environment, Scarcity, and Violence*. Princeton, NJ: Princeton University Press.

Homer-Dixon, T., Peluso, N. and Watts, M. (2003) 'Debating violent environments', *Environmental Change and Security Project Report*, 9: 89–96.

hooks, bell (1991) *Yearning: Race, Gender and Cultural Politics*. Boston, MA: South End Press.

Hosking, G. (2001) *Russia and the Russians: A History*. Cambridge, MA: Belknap Press.

Howarth, D. (2000) *Discourse*. Buckingham: Open University Press.

Hudson, B. (1977) 'The new geography and the new imperialism: 1870–1918', *Antipode*, 9(2): 12–19.

Huntington, S. (1991) *The Third Wave: Democratization in the Late Twentieth Century*. Norman and London: University of Oklahoma Press.

Huntington, S. (1996) *The Clash of Civilizations and the Remaking of World Order*. New York: Simon and Schuster.

Hyndman, J. (2000) *Managing Displacement*. Minneapolis: University of Minnesota Press.

Hyndman, J. (2001) 'Towards a feminist geopolitics', *The Canadian Geographer*, 45(2): 210–222.

Hyndman, J. (2003) 'Aid, conflict and migration: the Canada–Sri Lanka connection', *The Canadian Geographer*, 47(3): 251–268.

Hyndman, J. (2004a) 'Mind the gap: bridging feminist and political geography through geopolitics', *Political Geography*, 23(3): 307–322.

Hyndman, J. (2004b) 'The (geo)politics of gendered mobility', in L. Staeheli, E. Kofman and L. Peake (eds), *Mapping Women, Making Politics: Feminist Perspectives on Political Geography*. London and New York: Routledge. pp. 169–184.

Hyndman, J. (2005) 'Migration wars: refuge or refusal?' *Geoforum*, 36(1): 3–6.

Hyndman, J. (2007) 'Feminist geopolitics revisited: body counts in Iraq', *Professional Geographer*, 59(1): 35–46.

Hyndman, J. and Mountz, A. (2006) 'Refuge or refusal: the geography of exclusion', in D. Gregory and A. Pred (eds), *Violent Geographies*. London and New York: Routledge. pp. 77–92.

ICBL (2006) *Landmine Monitor Report: Toward a Mine-free World 2006*. Ottawa: International Campaign to Ban Landmines.

ICG (2007) *Afghanistan's Endangered Compact*. Kabul/Brussels: International Crisis Group.

Ignatieff, M. (2003) *Empire Lite: Nation-building in Bosnia, Kosovo, and Afghanistan*. New York: Vintage.

IHT (1999) 'To Paris, U.S. looks like a "hyperpower"', *International Herald Tribune, 5 February*. Retrieved 14 December 2007 from: http://www.iht.com/articles/1999/02/05/france.t_0.php

Independent International Commission on Kosovo (2000) *The Kosovo Report: Conflict, International Response, Lessons Learned*. New York: Oxford University Press.

Isin, E. (2002) *Being Political: Genealogies of Citizenship*. Minneapolis: University of Minnesota Press.

Jacobs, J. (1996) *Edge of Empire: Postcolonialism and the City*. London: Routledge.

Jacobs, L. and Shapiro, R. (2000) *Politicians Don't Pander: Political Manipulation and the Loss of Democratic Responsiveness*. Chicago: University of Chicago Press.

Jarosz, L. (2003) 'A human geographer's response to "Guns, Germs, and Steel": the case of agrarian development and change in Madagascar', *Antipode*, 35(4): 823–828.

Jefferson West, W. (2006) 'Religion as dissident geopolitics? Geopolitical discussions within the recent publications of Fethullah Gülen', *Geopolitics*, 11(2): 280–299.

Jeffries, F. (2001) 'Zapatismo and the intergalactic age', in Roger Burbach (ed.), *Globalization and Postmodern Politics: From Zapatistas to High Tech Robber Barons*. London: Pluto Press. pp. 129–144.

Jessop, B. (1990) *State Theory: Putting Capitalist States in their Place*. Cambridge: Polity Press.

Jessop, B. (1995) 'The regulation approach, governance and post-Fordisms: alternative perspectives on economic and political change?' *Economy and Society*, 24(3): 307–333.

Jessop, B. (1997) 'A neo-Gramscian approach to the regulation of urban regimes: accumulation strategies, hegemonic projects, and governance', in M. Lauria (ed.), *Reconstructing Urban Regime Theory: Regulating Urban Politics in a Global Economy*. Thousand Oaks, CA: Sage. pp. 51–73.

Jessop, B. (2002) 'Time and space in the globalization of capital and their implications for state power', *Rethinking Marxism*, 14(1): 97–117.

Jessop, B. (2004) *Hollowing out the 'Nation-state' and Multilevel Governance*. Cheltenham: Edward Elgar.

Johnson, C. (2000) *Blowback: The Costs and Consequences of American Empire*. New York: Henry Holt.

Johnston, N. (1995) 'Cast in stone: monuments, geography, and nationalism', *Environment and Planning D: Society and Space*, 13: 51–65.

Johnston, R. (2002) 'Manipulating maps and winning elections: measuring the impact of malapportionment and gerrymandering', *Political Geography*, 21(1): 1–31.

Jones, E. (1981) *The European Miracle: Environments, Economics and Geopolitics in the History of Europe and Asia*. Cambridge: Cambridge University Press.

Jones, M. (2001) 'The regional state and economic regulation: "partnerships for prosperity" or new scales of state power?' *Environment and Planning A*, 33: 1185–1211.

Jones, R., Goodwin, M., Jones, M. and Simpson, G. (2004a) 'Devolution, state personnel, and the production of new territories of governance in the United Kingdom', *Environment and Planning A*, 36: 89–109.

Jones, M., Jones, R. and Woods, M. (2004b) *An Introduction to Political Geography: Space, Place and Politics*. New York and London: Routledge.

Jones, M. and MacLeod, G. (2004) 'Regional space, spaces of regionalism: territory, insurgent politics and the English question', *Transactions of the Institute of British Geographers*, 29(4): 433–452.

Judah, T. (2008) *Kosovo: What Everyone Needs to Know*. New York: Oxford University Press.

Judt, T. (2005) *Postwar*. New York: Penguin Press.

Kaldor, M. (1999) *New and Old Wars: Organized Violence in a Global Era*. Cambridge: Polity Press.

Kaplan, C., Alarcón, N. and Moallem, M. (eds) (1999) *Between Woman and Nation: Nationalisms, Transnational Feminisms and the State*. Durham, NC and London: Duke University Press.

Kaplan, R. (1994) 'The coming anarchy: how scarcity, crime, overpopulation, and disease are rapidly destroying the social fabric of our planet', *Atlantic Monthly*, 273(2): 44–77.

Kearney, M. (1991) 'Borders and boundaries of state and self at the end of empire', *Journal of Historical Sociology*, 4(1): 52–74.

Kearns, G. (1993) '*Fin de siècle* geopolitics: Mackinder, Hobson and theories of global closure', in P. Taylor (ed.), *Political Geography of the Twentieth Century: A Global Analysis*. London: Belhaven Press. pp. 9–30.

Kearns, G. (2004) 'The political pivot of geography', *Geographical Journal*, 170: 337–346.

Keating, M. (1997) 'The invention of regions: political restructuring and territorial government in Western Europe', *Environment and Planning C: Government and Policy*, 15(4): 383–398.

Keith, M. and Pile, S. (1993) *Place and the Politics of Identity*. London: Routledge.

Kelly, P.F. (1999) 'The geographies and politics of globalization', *Progress in Human Geography*, 23: 379–400.

Kennan, G. (1947) 'The sources of Soviet conduct', *Foreign Affairs*, 25: 566–582.

Kent, R. (1993) 'Geographical dimensions of the Shining Path insurgency in Peru', *Geographical Review*, 83(4): 441–454.

King, A. (2003) 'Cultures and spaces of postcolonial knowledges', in K. Anderson, M. Domosh, S. Pile and N. Thrift (eds), *Handbook of Cultural Geography*. London: Sage. pp. 381–397.

Kirby, A. (1997) 'Is the state our enemy?' *Political Geography*, 16(1): 1–12.

Kliot, N. and Charney, I. (2006) 'The geography of suicide terrorism in Israel', *GeoJournal*, 66: 353–373.

Kobayashi, A. and Peake, L. (1994) 'Unnatural discourse: "race" and gender in geography', *Gender, Place and Culture,* 1(2): 225–243.

Kofman, E. (1996) 'Feminism, gender relations and geopolitics: problematic closures and opening strategies', in E. Kofman and G. Youngs (eds), *Globalization: Theory and Practice*. London and New York: Pinter. pp. 209– 224.

Krasner, S. D. (1999) *Sovereignty: Organized Hypocrisy*. Princeton, NJ: Princeton University Press.

Kristeva, J. (1982) *The Powers of Horror*. New York: Columbia University Press.

Kropotkin, Peter. (1885) 'What geography should be', *Nineteenth Century*, 18: 940–956.

Kuchukeeva, A. and O'Loughlin, J. (2003) 'Civic engagement and democratic consolidation in Kyrgyztan', *Eurasian Geography and Economics*, 44(8): 557–587.

Kyle, D. and Koslowski, Rey (eds) (2001) *Global Human Smuggling: Comparative Perspectives*. Baltimore, MD: Johns Hopkins University Press.

Laborde, C. (2000) 'The concept of the state in British and French political thought', *Political Studies*, 48: 540–557.

Lacan, J. (1977) *Ecrits: A Selection*. New York: Norton.

Laclau, E. and Mouffe, C. (1985) *Hegemony and Socialist Strategy: Towards a Radical Democratic Politics*. London: Verso.

Landes, D. (1998) *The Wealth and Poverty of Nations: Why Some Are So Rich and Some So Poor*. New York: W.W. Norton and Co.

Laqueur, W. (1999) *The New Terrorism*. Oxford: Oxford University Press.

Latham, A. and McCormack, D. (2004) 'Moving cities: rethinking the materialities of urban geographies', *Progress in Human Geography,* 28: 701–724.

Laurie, N. (ed.) (2007) 'Pro-poor water? The privatisation and global poverty debate' (Special themed issue), *Geoforum*, 38(5): 753–907.

Lawson, V. (1999) 'Questions of migration and belonging: understandings of migration under neoliberalism in Ecuador', *International Journal of Population Geography,* 5(4): 261–276.

Le Billion, P. (2001) 'The political ecology of war: natural resources and armed conflicts', *Political Geography*, 20(5): 561–584.

Le Billon, P. (2004) 'The geopolitical economy of "resource wars"', *Geopolitics,* 9(1): 1–28.

Le Billon, P. (2007) 'Geographies of war: perspectives on "resource wars"', *Geography Compass,* 1(2): 163–182.

Le Billon, P. and El Khatib, F. (2004) 'From free oil to "freedom oil"? Terrorism, war and US geopolitics in the Persian Gulf', *Geopolitics,* 9(1): 109–137.

Leitner, H. and Miller, B. (2007) 'Scale and the limitations of ontological debate: a commentary on Marston, Jones and Woodward', *Transactions of the Institute of British Geographers,* 32(1): 116–125.

Lemon, A. (2007) 'Perspectives on democratic consolidation in southern Africa: the five general elections of 2004', *Political Geography,* 26(7): 824–850.

Lester, A. (2001) *Imperial Networks: Creating Identities in Nineteenth-century South Africa and Britain*. London: Routledge.

Levitas, D. (2002) *The Terrorist Next Door: The Militia Movement and the Radical Right*. New York: Thomas Dunne Books.

Ley, D. (1996) *The New Middle Class and the Remaking of the Central City*. Oxford: Oxford University Press.

Ley, D. and Hiebert, D. (2001) 'Immigration policy as population policy', *The Canadian Geographer/Le Géographe Canadien*, 45(1): 120–125.

Ley, D. and Kobayashi, A. (2005) 'Back to Hong Kong: return migration or transnational sojourn?' *Global Networks*, 5(2), 111–127.

Liao, S.-H. (2005) 'Will China become a military space superpower?' *Space Policy*, 21(3): 205–212.

Lindner, P. (2007) 'Localising privatisation, disconnecting locales – mechanisms of disintegration in post-socialist rural Russia', *Geoforum*, 38(3): 494–504.

Lintz, G., Müller, B. and Schmude, K. (2007) 'The future of industrial cities and regions in central and eastern Europe', *Geoforum*, 38(3): 512–519.

Lipietz, A. (1992) *Towards a New Economic Order: Postfordism, Ecology, and Democracy*, translated from the French by Malcolm Slater. New York: Oxford University Press.

Livingstone, D. (1992) *The Geographical Tradition*. London: Blackwell.

Lombardi, M. J. (2005) 'The decline of the American superpower', *Defense and Security Analysis*, 21(3): 312–321.

Long, J. (2006) 'Border anxiety in Israel–Palestine', *Antipode*, 38(1): 107–127.

Loomba, A. (2005) *Colonialism/postcolonialism*. New York: Routledge.

Loughlin, J. (1996) '"Europe of the regions" and the federalization of Europe', *Publius*, 26(4): 141–162.

Luke, T. (1996) 'Governmentality and contragovernmentality: rethinking sovereignty and territoriality after the Cold War', *Political Geography*, 15(6/7): 491–507.

Luke, T. (1997) 'Is the state our enemy? Or is our state the enemy?: A reply to Kirby', *Political Geography* 16(1): 21–26.

McClintock, A. (1992) 'The angel of progress: pitfalls of the term "post-colonialism"', *Social Text*, 31/32: 84–98.

McColl, R.W. (1969) 'The insurgent state: territorial bases of revolution', *Annals of the Association of American Geographers*, 59: 613–631.

McDowell, L. (1999) *Gender, Identity and Place: Understanding Feminist Geographies*. Minneapolis: University of Minnesota Press.

McEwan, C. (2000) 'Engendering citizenship: gendered spaces of democracy in South Africa', *Political Geography*, 19(5): 627–651.

McGee, T.G. (1974) 'In praise of tradition: towards a geography of anti-development', *Antipode*, 6(3): 30–47.

McGreal, C. (2004) 'Deadly thirst', *Guardian*, 13 January. Retrieved 1 November 2007 from http://www.guardian.co.uk/environment/2004/jan/13/water.israel

McHugh, K. (2000) 'Inside, outside, upside down, backward, forward, round and round: a case for ethnographic study in migration', *Progress in Human Geography*, 24(1): 71–89.

McHugh, K. and Mings, R. (1996) 'The circle of migration: attachment to place in ageing', *Annals of the Association of American Geographers*, 86: 530–550.

Mackinder, H. J. (1904) 'The geographical pivot of history', *The Geographical Journal,* 23(4): 421–437.

McLafferty, S. and Preston, V. (1991) 'Gender, race, and commuting among service sector workers', *Professional Geographer,* 43(1): 1–15.

McLaughlin Mitchell, S. (ed.) (2006) Special issue: 'Conflict and cooperation over international rivers', *Political Geography*, 25(4): 357–477.

Mahan, A. T. (1890) *The Influence of Sea Power upon History, 1660–1783.* Boston, MA: Little, Brown.

Mamadouh, V. (2003) 'Some notes on the politics of political geography', *Political Geography,* 22: 663–675.

Mann, M. (1999) 'The dark side of democracy: the modern tradition of ethnic and political cleansing', *New Left Review,* 235: 18–45.

Manzo, K. A. (1996) *Creating Boundaries: The Politics of Race and Nation*. Boulder, CO: Lynne Rienner.

Marchand, M.H. and Runyan, A.S. (eds) (2000) *Gender and Global Restructuring: Sightings, Sites and Resistances.* London and New York: Routledge.

Mares, P. (2002) *Borderline.* Sydney: UNSW Press.

Marston, S. A. (2000) 'The social construction of scale', *Progress in Human Geography,* 24(2): 219–242.

Marston, S. A. (2002) 'Making difference: conflict over Irish identity in the New York City St Patrick's Day Parade', *Political Geography*, 21: 373–392.

Marston, S. A., Jones III, J. P. and Woodward, K. (2005) 'Human geography without scale', *Transactions of the Institute of British Geographers,* 30(4): 416–432.

Marx, K. (1961) *Economic and Philosophic Manuscripts of 1844.* Moscow: Foreign Languages Publishing House.

Marx, K. (1962) *Capital Volumes 1–3*, translated from the third German edition by Samuel Moore and Edward Aveling and edited by Frederick Engels. Moscow: Foreign Languages Publishing House.

Marx, K. and Engels, F. (1848) *The Communist Manifesto.* Retrieved 13 January, 2008, from http://www.marxists.org/archive/marx/works/1848/communist-manifesto/index.htm

Massey, D. (1989) 'Forword', in R. Peet and N. Thrift (eds), *New Models in Geography*. London: Unwin Hyman. pp. IX–XI.

Massey, D. (1993) 'Power-geometry and a progressive sense of place', in J. Bird, B. Curtis, T. Putnam, G. Robertson and L. Tickner (eds), *Mapping the Futures: Local Cultures, Global Change.* New York: Routledge. pp. 59–69.

Massey, D. (1994) *Space, Place, and Gender*. London and New York: Routledge.

Mayer, T. (2004) 'Embodied nationalisms', in L. Staeheli, E. Kofman and L. Peake (eds), *Mapping Women, Making Politics: Feminist Perspectives on Political Geography*. London and New York: Routledge. pp. 153–168.

Mbembe, A. (2000) 'At the edge of the world: boundaries, territoriality, and sovereignty in Africa', *Public Culture,* 12(1): 259–284.

Mercer, C. (2002) 'NGOs, civil society and democratization: a critical review of the literature', *Progress in Development Studies*, 2: 5–22.

Miller, B. (2004) 'Spaces of mobilization: transnational social movements', in C. Barnett and M. Low (eds), *Spaces of Democracy*. London: Sage. pp. 223–246.

Mitchell, D. (2003) *The Right to the City: Social Justice and the Fight for Public Space.* New York: Guilford Press.

Mitchell, J. (2003) 'Urban vulnerability to terrorism as hazard', in S. Cutter, D. Richardson and T. Wilbanks (eds), *The Geographical Dimensions of Terrorism*. New York: Routledge. pp. 17–26.

Miyares, I. (1997) 'Changing perceptions of space and place as measures of Hmong acculturation', *Professional Geographer*, 49(2): 214–224.

Mohanty, C. (2003) *Feminism without Borders: Decolonizing Theory, Practicing Solidarity*. Durham, NC: Duke University Press.

Monbiot, G. (2000) *Captive State: The Corporate Takeover of Britain*. London: Macmillan.

Monbiot, G. (2005) 'A truckload of nonsense', *The Guardian*, 4 June. Retrieved 18 September 2008 from http://www.guardian.co.uk/comment/story/0,,1505816,00.html

Monk, J. and Hanson, S. (1982) 'On not excluding half of the human in human geography', *Professional Geographer,* 34(1): 11–23.

Morrill, R. (2004) 'Representation, law and redistricting in the United States', in C. Barnett and M. Low (eds), *Spaces of Democracy*. London: Sage. pp. 67–92.

Morrill, R., Knopp, L. and Brown, M. (2007) 'Anomalies in red and blue: exceptionalism in American electoral geography', *Political Geography*, 26(5): 525–553.

Mouffe, C. (1995) 'Post-marxism: democracy and identity', *Environment and Planning D: Society and Space*, 13: 259–265.

Mountz, A. (1995) '*Daily life in a transnational migrant community: the fusion of San Agustín, Oaxaca and Poughkeepsie, New York*'. Unpublished Master's thesis, Dartmouth College, Hanover, New Hampshire.

Mountz, A. (2004) Embodying the nation-state: Canada's response to human smuggling. *Political Geography*, 23: 323–345.

Mountz, A. (2006) 'Human smuggling and the Canadian state', *Canadian Foreign Policy*, 13(1): 59–80.

Mountz, A. (2009) *States of Migration: Human Smuggling and the Borders of Sovereignty*. Mineapolis: University of Minnesotta Press.

Mountz, A. (forthcoming) 'Stateless by geographical design', *International Migration Review*.

Mountz, A. and Hyndman, J. (2006) 'Feminist approaches to the global intimate', *Women's Studies Quarterly*, 34(1 & 2): 446–463.

Mountz, A. and Wright, R. (1996) 'Daily life in the transnational migrant community of San Agustín, Oaxaca and Poughkeepsie, New York', *Diaspora,* 6: 403–428.

Mountz, A. Miyares, I., Wright, R. and Bailey, A. (2002) 'Methodologically becoming: power, knowledge and team research', *Gender, Place and Culture*, 10(1): 29–46.

Mueller, J. (2007) 'Reacting to terrorism: probabilities, consequences, and the persistence of fear'. Paper presented at the International Studies Association Meeting, Chicago Illinois, 26 February–4 March.

Mullings, B. (1999) 'Sides of the same coin? Coping and resistance among Jamaican data-entry operators', *Annals of the Association of American Geographers*, 89(2): 290–311.

Murphy, A. (1996) 'The sovereign state system as political-territorial ideal: historical and contemporary considerations', in T. J. Biersteker and C. Weber (eds), *State Sovereignty as Social Construct*. Cambridge: Cambridge University Press. pp. 81–120.

Murphy, A., Bassin, M., Newman, D., Reuber, P. and Agnew, J. A. (2004) 'Is there a politics to geopolitics?' *Progress in Human Geography, 28*(5): 619–640.

Murray, C. (1994) '*Losing Ground: American Social Policy, 1950–1980.* New York: Basic Books.

Mutibwa, P. (1992) *Uganda since Independence: A Story of Unfulfilled Hopes.* Trenton, NJ: Africa World Press.

Myers, G. (2003) *Verandahs of Power: Colonialism and Space in Urban Africa.* Syracuse: Syracuse University Press.

Myers, S.L. (2004) 'Putin issues plan to tighten grasp, citing terrorism: overhaul of political system – opponents call it step back', *New York Times*, 14 September: A1, A14.

Nagar, R., Lawson, V., McDowell, L. and Hanson, S. (2002) 'Locating globalization: feminist (re)readings of the subjects and spaces of globalization', *Economic Geography*, 78(3): 257–284.

Nagel, C. (2001) Review essay. *Political Geography*, 20(2): 247–256.

Nagengast, C. (1994) 'Violence, terror, and the crisis of the state', *Annual Review of Anthropology*, 23: 109–136.

Naím, M. (2000) 'Washington consensus or Washington confusion?' *Foreign Policy*, 118: 87–103.

Nash, C. (1999) 'Irish placenames: post-colonial locations', *Transactions of the Institute of British Geographers*, 24(4): 457–480.

National Security Strategy of the United States of America (2006) Washington, DC: White House.

Nelson, D. (1999) *A Finger in the Wound.* Durham, NC: Duke University Press.

Nevins, J. (2002) *Operation Gatekeeper: The Rise of the 'Illegal Alien' and the Remaking of the US–Mexico Boundary.* New York: Routledge.

NIC (2004) *Mapping the Global Future: Report of the National Intelligence Council's 2020 Project.* Washington, DC: National Intelligence Council.

NRC (2004) *Profile of Internal Displacement: Serbia and Montenegro.* Geneva: Norwegian Refugee Council/Global IDP Project.

Office of the Secretary of Defense (2007) *Annual Report to Congress: Military Power of the People's Republic of China 2007.* Washington, DC: Department of Defense.

Ohmae, K. (1995) *The End of the Nation-state.* London: HarperCollins.

Olivera, O. (2004) *¡Cochabamba! Water War in Bolivia!* Boston, MA: South End Press.

O'Loughlin, J. (2004) 'Global democratization', in C. Barnett and M. Low (eds), *Spaces of Democracy.* London: Sage. pp. 23–44.

O'Loughlin, J. (2005a) 'The war on terrorism, academic publication norms, and replication', *The Professional Geographer*, 57(4): 588–591.

O'Loughlin, J. (2005b) 'The political geography of conflict: civil wars in the hegemonic shadow', in C. Flint (ed.), *The Geography of War and Peace.* Oxford: Oxford University Press. pp. 85–110.

O'Loughlin, J. and Raleigh, C. (2007) 'Spatial analysis of civil war violence', in K. Cox, M. Low and J. Robinson (eds), *A Handbook of Political Geography.* London: Sage.

Ong, A. (1999) *Flexible Citizenship.* Durham, NC: Duke University Press.

Ong, A. (2006) *Neoliberalism as Exception: Mutations in Citizenship and Sovereignty.* Durham, NC: Duke University Press.

Onnepalu, T. (2000) *Border State.* Evanston, IL: Northwestern University Press.

Otis, G. (2002) A World without Water, 20 August Retrieved 1 November 2007 from http://www.globalpolicy.org/ngos/role/policymk/conf/2002/0827water.htm

Ó Tuathail, G. (1986) 'The language and nature of the "new" geopolitics: the case of US–El Salvador relations', *Political Geography Quarterly*, 5: 73–85.

Ó Tuathail, G. (1996a) *Critical Geopolitics: The Politics of Writing Global Space*. Minneapolis: University of Minnesota Press.

Ó Tuathail, G. (1996b) 'An anti-geopolitical eye? Maggie O'Kane in Bosnia, 1992–94', *Gender, Place and Culture*, 3(2): 171–185.

Ó Tuathail, G. (1996c) *Critical Geopolitics: The Politics of Writing Global Space*. Minneapolis: University of Minnesota Press.

Ó Tuathail, G. (2006) 'Thinking critically about geopolitics', in G. Ó Tuathail, S. Dalby and P. Routledge (eds), *The Geopolitics Reader*. New York: Routledge.

Ó Tuathail, G. and Dalby, S. (1998) 'Introduction: Rethinking geopolitics: towards a critical geopolitics', in G.Ó Tuathail and S. Dalby (eds), *Rethinking Geopolitics*. New York: Routledge, pp. 1–15.

Ould-Mey, M. (1996) *Global Restructuring and Peripheral States: The Carrot and the Stick in Mauritania*. Lanham, MD: Littlefield Adams Books.

Paasi, A. (2001) 'Europe as a social process and discourse: considerations of place, boundaries and identity', *European Urban and Regional Studies*, 8(1): 7–28.

Paasi, A. (2003) 'Region and place: regional identity in question', *Progress in Human Geography*, 27(4): 475–485.

Painter, J. (1995) *Politics, Geography and 'Political Geography': A Critical Perspective*. London: Arnold.

Parker, G. (2004) *Sovereign City: The City-state through History*. London: Reaktion Books.

Pavlínek, P. and Pickles, J. (2000) *Environmental Transitions: Transformation and Ecological Defence in Central and Eastern Europe*. London and New York: Routledge.

Pavlovskaya, M. and Hanson, S. (2001) 'Privatization of the urban fabric: gender and local geographies in downtown Moscow', *Urban Geography*, 22(1): 4–28.

Peck, J. (2002) 'Political economies of scale: fast policy, interscalar relations, and neoliberal workfare', *Economic Geography*, 78(3): 331–360.

Peck J. (2006) 'Liberating the city: between New York and New Orleans', *Urban Geography*, 27: 681–713.

Peet, R. (1985) 'The social origins of environmental determinism', *Annals of the Association of American Geographers*, 75: 309–333.

Peet, R. (1998) *Modern Geographical Thought*. Oxford: Blackwell.

Peet, R. (2002) 'Ideology, discourse, and the geography of hegemony: from socialist to neoliberal development in postapartheid South Africa', *Antipode*, 34(1): 54–84.

Peet, R. and Thrift, N. (1989) 'Political economy and human geography', in R. Peet and N. Thrift (eds), *New Models in Geography*. London: Unwin Human. pp. 3–29.

Peluso, N. and Watts, M. (eds) (2001) *Violent Environments*. Ithaca, NY: Cornell University Press.

Perreault, T. (2006) 'From the guerra del agua to the guerra del gas: resource governance, neoliberalism and popular protest in Bolivia', *Antipode*, 38: 150–172.

Peters, E. J. (1997) 'Challenging the geographies of "Indianess": the Batchewana case', *Urban Geography*, 18(1): 56–62.

Peters, E. J. (1998) 'Subversive spaces: First Nations women and the city in Canada', *Environment and Planning D: Society and Space*, 16(6): 665–686.

Pickles, J. (1995) 'Representations in an electronic age: geography, GIS, and democracy', in *Ground Truth: The Social Implications of Geographic Information Systems*. New York: Guilford Press. pp. 1–3.

Pickles, J. (2000) 'Ethnicity, violence and the production of regions', *Growth and Change*, 31: 139–150.

Power, M. and Sidaway, J. (2004) 'The degeneration of tropical geography', *Annals of the Association of American Geographers*, 94(3): 585–601.

Pratt, G. (2000) Feminist Geographies. In R. J Johnston, D. Gregory, G. Pratt, and M. Watts (eds), *The Dictionary of Human Geography* (4th edn). Oxford: Blackwell Publishers. pp. 259–262.

Pratt, G. (2004) *Working Feminism*. Philadelphia: Temple University Press.

Pratt, G. and the Philippine Women Centre, Vancouver Canada (1998) 'Inscribing domestic work on Filipina bodies', in H. Nast and S. Pile (eds), *Places through the Body*. New York: Routledge. (pp. 211–226).

Pratt, M.B. (2003) *The Dirt She Ate*. Pittsburgh: University of Pittsburgh Press.

Pred, A. (1997) 'Re-presenting the extended present moment of danger', in G. Benko and U. Strohmayer (eds), *Space and Social Theory*. Oxford: Blackwell. pp. 117–140.

Prytherch, D. L. (2007) 'Urban geography with scale: rethinking scale via Wal-Mart's "geography of big things"', Urban Geography, 28(5): 456–482.

Puar, J. (2007) *Terrorist Assemblages: Homonationalism in Queer Times*. Durham, NC: Duke University Press.

Pudup, M. B. (1988) 'Arguments within regional geography', *Progress in Human Geography,* 12(3): 369–390.

Purcell, M. and Nevins, J. (2005) 'Pushing the boundary: State restructuring, state theory, and the case of U.S.–Mexico border enforcement in the 1990s', *Political Geography*, 24: 211–235.

Quart, A. (2008) 'When girls will be boys', *New York Times Magazine,* 16 March: 32–37.

Quattrochi, D. A. and Goodchild, M. F. (1997) *Scale in Remote Sensing and GIS*. Boca Raton, FL: Lewis Publishers.

Quigley, C. (1995) *Tragedy and Hope: A History of the World in our Time*. New York: GSG & Associates.

Raco, M. and Imrie, R. (2000) 'Governmentality and rights and responsibilities in urban policy', *Environment and Planning A* 32, 2187–2204.

Radcliffe, S. and Westwood, S. (1993) *Remaking the Nation: Place, Identity and Politics in Latin America*. London and New York: Routledge.

Ramutsindela, M. F. (2001) 'Down the post-colonial road: reconstructing the post-apartheid state in South Africa', *Political Geography*, 20(1): 57–84.

Rangan, H. (2001) *Of Myths and Movements: Rewriting Chipko into Himalayan History*. London: Verso.

Rangan, H. and Gilmartin, M. (2002) 'Gender, traditional authority, and the politics of rural reform in South Africa', *Development and Change*, 33(4): 633–658.

Reclus, Élisée (1991) *L'Homme et la terre*: *Histoire contemporaine*, Paris: Fayard.

Richards, P. (2001) 'Are "forest" wars in Africa resource conflicts? The case of Sierra Leone', in N. Peluso and M. Watts (eds), *Violent Environments*. Ithaca, NY: Cornell University Press. pp. 65–84.

Robbins, P. (2004) *Political Ecology: A Critical Introduction*. Malden, MA: Blackwell.

Roberts, S. M. (2004) 'Gendered globalization', in L. Staeheli, E. Kofman and L. Peake (eds), *Mapping Women, Making Politics*. London: Routledge. pp. 127–140.

Robinson, J. (2003a) 'Postcolonialising geography: tactics and pitfalls', *Singapore Journal of Tropical Geography*, 24(3): 273–289.

Robinson, J. (2003b) 'Political geography in a postcolonial context', *Political Geography*, 22: 647–651.

Rogerson, C. (2000) 'Local economic development in an era of globalisation: the case of South African cities', *Tijdschrift voor Economische en Sociale Geografie*, 91(4): 397–411.

Rose, G. (1993) *Feminism and Geography: The Limits of Geographical Knowledge*. Cambridge: Polity Press.

Rose, G. (2001) *Visual Methodologies*. London: Sage.

Rostow, W. (1960) *The Stages of Economic Growth: A Non-communist Manifesto*. Cambridge: Cambridge University Press.

Routledge, P. (2003a) 'Anti-geopolitics', in J. Agnew, K. Mitchell and G. Toal (eds), *A Companion to Political Geography*. Malden, MA: Blackwell. pp. 236–248.

Routledge, P. (2003b) 'Convergence space: process geographies of grassroots globalization networks', *Transactions of the Institute of British Geographers*, 28(3): 333–349.

Roy, A. (1999) *The Cost of Living*. London: Flamingo.

Rubin, B. R. (2002) *The Fragmentation of Afghanistan: State Formation and Collapse in the International System*. New Haven, CT: Yale University Press.

Sachs, J. (2005) *The End of Poverty: Economic Possibilities for our Time*. New York: Penguin Press.

Sack, R. D. (1986) *Human Territoriality: Its Theory and History*. New York: Cambridge University Press.

Said, E. (1979) *Orientalism*. New York: Pantheon Books.

Said, E. (1993) *Culture and Imperialism*. New York: Vintage.

Saltmarsh, R. (2001) 'Journey into autobiography: a coalminer's daughter', in P. Moss (ed.), *Placing Autobiography in Geography*. Syracuse, NY: Syracuse University Press.

Sangtin Writers Collective and Nagar, R. (2006) *Playing with Fire*. Minneapolis: University of Minnesota Press.

Sassen, S. (2006) *Territory, Authority, Rights: From Medieval to Global Assemblages*. Princeton, NJ: Princeton University Press.

Schaefer, F. (1953) 'Exceptionalism in geography: a methodological examination', *Annals of the Association of American Geographers*, 43: 226–249.

Schivelbusch, W. (2004) *The Culture of Defeat: On National Trauma, Mourning, and Recovery*. New York: Metropolitan Books.

Schoenberger, E. (2004) 'The Spatial Fix revisited', *Antipode*, 36(3): 427–433.

Schumpeter, J.A. (1992) *Capitalism, Socialism and Democracy* (5th edn). London and New York: Routledge.

Scott, A. J. (1998). *Regions and the World Economy: The Coming Shape of Global Production, Competition, and Political Order*. New York: Oxford University Press.

Scott, J. (1998) *Seeing like a State: How Certain Schemes to Improve the Human Condition Have Failed*. New Haven and London: Yale University Press.

Secor, A. (2001) 'Ideologies in crisis: political cleavages and electoral politics in Turkey in the 1990s', *Political Geography*, 20: 539–560.

Secor, A. J. (2002) 'The veil and urban space in Istanbul: Women's dress, mobility and Islamic knowledge', *Gender, Place and Culture*, 9(1): 5–22.

Secor, A. (2004) 'Feminist electoral geography', in L. Staeheli, E. Kofman and L. Peake (eds), *Mapping Women, Making Politics: Feminist Perspectives on Political Geography*. London and New York: Routledge. pp. 261–272.

Sharp, J. (1996) 'Gendering nationhood: a feminist engagement with national identity', in N. Duncan (ed.), *Bodyspace: Destabilising Geographies of Gender and Sexuality*. New York: Routledge. pp. 97–108.

Sharp, J. (2000) *Condensing the Cold War: 'Reader's Digest' and American Identity*. Minneapolis: University of Minnesota Press.

Sharp, J. (2003) 'Feminist and postcolonial engagements', in J. Agnew, K. Mitchell and G. Toal (eds), *A Companion to Political Geography*. Malden, MA: Blackwell. pp. 58–74.

Sharp, J. (2007) Geography and Gender: Finding Feminist Political Geographies. *Progress in Human Geography*, 31(3): 381–387.

Sharpe, J. (2004) 'Doing feminist political geographies', in L. Staeheli, E. Kofman and L. Peake (eds), *Mapping Women, Making Politics: Feminist Perspectives on Political Geography*. London and New York: Routledge. pp. 87–98.

Shepherd, E. (2002) 'The spaces and times of globalization: place, scale, networks, and positionality', *Economic Geography,* 78(3): 307–330.

Shirlow, P. (2006) 'Belfast: the "post-conflict" city', *Space and Polity*, 10(2): 99–107.

Shirlow P. and Murtagh, B. (2006) *Belfast: Segregation, Violence and the City*. London: Pluto Press.

Shiva, V. (2002) *Water Wars: Privatization, Pollution and Profit*. London: Pluto Press.

Shohat, E. (2001) 'Area studies, transnationalism, and the feminist production of knowledge', *Signs*, 26(4): 1269–1272.

Short, J. R. (1993) *An Introduction to Political Geography*, 2nd edn. London and New York: Routledge.

Sibley, D. (1995) *Geographies of Exclusion*. London: Routledge.

Sidaway, J. (2000) 'Postcolonial geographies: an exploratory essay', *Progress in Human Geography*, 24(4): 591–612.

Silke, L. (1973) 'Arab brinkmanship: output pledge may signal recognition of potential risk', *New York Times*, 23 December: 65.

Silvey, R. (2004) 'Power, difference and mobility: feminist advances in migration studies', *Progress in Human Geography*, 28(4): 490–506.

Silvey, R. and Lawson, V. (1999) 'Placing the migrant', *Annals of the Association of American Geographers*, 89(1): 121–132.

Sinclair, A. and Byers, M. (2007) 'When US scholars speak of "sovereignty," what do they mean?' *Political Studies,* 55(2): 318–340.

Slater, D. (1973) 'Geography and underdevelopment – I', *Antipode*, 5(3): 21–32.

Slater, D. (1977) 'Geography and underdevelopment – II', *Antipode*, 9(3): 1–31.

Slater, D. (2004) *Geopolitics and the Post-colonial: Rethinking North–South Relations*. Malden, MA: Blackwell.

Smismans, S. (2006). *New Modes of Governance and the Participatory Myth*. Retrieved 23 November 2007 from European University Institute, Robert Schuman Centre for Advanced Studies Web site, http://www.eu-newgov.org/database/DELIV/D11D05_NMG_and_the_participatory_myth.pdf

Smith, A. (1957) *Selections from 'The Wealth of Nations'*, ed. George Stigler. Arlington Heights, IL: AHM.

Smith, A. and Pickles, J. (eds) (1998) *Theorizing Transition: The Political Economy of Post-communist Transformations*. New York: Routledge.

Smith, A. and Stenning, A. (2006) 'Beyond household economies: articulations and spaces of economic practice in postsocialism', *Progress in Human Geography*, 30(2): 190–213.

Smith, A. M. (1998) *LaClau and Mouffe: The Radical Democratic Imaginary*. New York: Routledge.

Smith, M. P. and Guarnizo, L. (eds) (1998) *Transnationalism from Below*. Transaction Publishers.

Smith, N. (1984) *Uneven Development: Nature, Capital and the Production of Space*. Oxford: Basil Blackwell.

Smith, N. (1987) 'Dangers of the empirical turn: some comments on the CURS Initiative', *Antipode*, 19: 59–68.

Smith, N. (1992) 'Contours of a spatialized politics: homeless vehicles and the production of geographical space', *Social Text*, 33: 54–81.

Smith, N. (1996) *The New Urban Frontier: Gentrification and the Revanchist City*. New York: Routledge.

Smith, N. (2003) *American Empire: Roosevelt's Geographer and the Prelude to Globalization*. California: University of California Press.

Smith, N. (2005a) 'Neo-critical geography, or, the flat pluralist world of business class', *Antipode*, 37(5): 887–899.

Smith, N. (2005b) *The Endgame of Globalization*. New York: Routledge.

Soja, E. (1989) *Postmodern Geographies: The Reassertion of Space in Critical Social Theory*. London: Verso

Southern, J. (2000) 'Blue collar, white collar: deconstructing classification', in J. K. Gibson-Graham, S. Resnick, and R. Wolff (eds), *Class and its Others*. Minneapolis: University of Minnesota Press. pp. 191–224.

Sparke, M. (2003) 'American empire and globalisation: postcolonial speculations on neocolonial enframing', *Singapore Journal of Tropical Geography*, 24(3): 373–389.

Sparke, M. (2005) *In the Space of Theory*. Minneapolis: University of Minnesota Press.

Sparke, M. (2006) 'The neoliberal nexus', *Political Geography*, 25(2): 151–180.

Sparke, M., Sidaway, J. Bunnell, T. and Grundy-Warr, C. (2004) 'Triangulating the borderless world: geographies of power in the India–Malaysia–Singapore growth triangle', *Transactions*, 29(4): 485–498.

Spronk, S. and Webber, J. (2007) 'Struggles against accumulation by dispossession in Bolivia: the political economy of natural resource contention', *Latin American Perspectives*, 34(2): 31–47.

Staeheli, L. and Kofman, E. (2004) 'Mapping gender, making politics', in L. Staeheli, E. Kofman, and L. Peake (eds), *Mapping Women, Making Politics: Feminist Perspectives on Political Geography*. London: Routledge. pp. 1–15.

Staeheli, L. and Mitchell, D. (2005) 'The complex politics of relevance in geography', *Annals of the Association of American Geographers*, 95(2): 357–372.

Staeheli, L., Kofman, E. and Peake, L. (eds) (2004) *Mapping Women, Making Politics: Feminist Perspectives on Political Geography*. New York and London: Routledge.

Stansfield, G. (2007) *Iraq: People, History, Politics* Cambridge and Malden, MA: Polity Press.

Staten, P. (1987) 'The economic climate for a banking crisis', in M. Staten (ed.), *The Embattled Farmer*. Golden, CO: Fulcrum. pp. 133–148.

Statistical Office of Kosovo (2008) *Demographic Changes of the Kosovo Population 1948–2006*. Prishtina: Government of Kosovo.

Steinberg, P. (1997) 'And are the anti-statist movements our friends?' *Political Geography*, 16(1): 13–19.

Steinberg, P. (2003) 'Hegemony, identity, the particular and the universal: comments on Joanne Sharp's "Condensing the cold war"', *Geopolitics*, 8(2): 166–174.

Stenning, A. (2000) 'Placing (post-)socialism: the making and remaking of Nowa Huta, Poland', *European Urban and Regional Studies*, 7(2): 99–118.

Stenning, A. (2003) 'Shaping the economic landscapes of post-socialism? Labour, workplace and community in Nowa Huta, Poland', *Antipode*, 35(4): 761–780.

Stenning, A. (2005) 'Post-socialism and the changing geographies of the everyday in Poland', *Transactions of the Institute of British Geographers*, 30(1): 113–127.

Steinberg, P., Walter, A. and Sherman-Morris, K. (2002) Using the Internet to Integrate Thematic and Regional Approaches in Geographic Education. *The Professional Geographer*, 54(3): 332–348.

Stiglitz, J. (2003) *Globalization and its Discontents*. New York: W.W. Norton.

Storey, D. (2001) *Territory: The Claiming of Space*. Harlow, Essex: Pearson Education.

Straussfogel, D. (1997) 'World-systems theory: toward a heuristic and pedagogic conceptual tool', *Economic Geography*, 73(1): 118–130.

Sundberg, J. (2003) 'Conservation and democratization: constituting citizenship in the Maya Biosphere Reserve, Guatemala', *Political Geography*, 22: 715–740.

Swyngedouw, E. (1997) 'Neither global nor local: globalisation and the politics of scale', in K. Cox (ed.), *Spaces of globalization*. New York: Guilford Press. pp. 137–166.

Tabor, J. (1997) 'A failed utopian vision', *Political Geography*, 16(1): 27–31.

Taylor, F. (1911) *The Principles of Scientific Management*. New York: Harper and Brothers.

Taylor, P. (1982) 'A materialist framework for political geography', *Transactions of the Institute of British Geographers*, 7(1): 15–34.

Taylor, P. (1993) *Political Geography: World-economy, Nation-state and Locality*, 3rd edn. Harlow: Longman.

Taylor, P. (1994) 'The state as container: territoriality in the modern world-system', *Progress in Human Geography*, 18(2): 151–162.

Taylor, P. (1996) *The Way the Modern World Works: World Hegemony to World Impasse*. Chichester: John Wiley and Sons.

Taylor, P. (2003) 'Radical political geographies', in J. Agnew, K. Mitchell and G. Toal (eds), *A Companion to Political Geography*. Massachusetts and Oxford: Blackwell. pp. 47–58.

Taylor, S. (2005) 'The Pacific solution or a Pacific nightmare?: The difference between burden shifting and responsibility sharing', *Asia-Pacific Law & Policy Journal*, 6(1): 1–43.

Teschke, B. (2003) *The Myth of 1648: Class, Geopolitics, and the Making of Modern International Relations*. New York: Verso.

Thomas, A., Crow, B., Frenz, P., Hewitt, T., Kassam, S. and Treagust, S. (1994) *Third World Atlas*, 2nd edn. Washington, DC: Taylor and Francis.

Thrift, N. (1996) *Spatial Formations*. London: Sage.

Tishkov, V. (2004) *Chechnya: Life in a War-torn Society*. Berkeley: University of California Press.

Toal, G. (2003) 'Re-asserting the regional: political geography and geopolitics in a world thinly known', *Political Geography*, 22: 653–655.

Todd, E. (2004) *After the Empire: The Breakdown of the American Order*. London: Constable.

Torpey, J. (2000) *The Invention of the Passport: Surveillance, Citizenship and the State*. Cambridge: Cambridge University Press.

Trevor-Roper, H. (1983) 'The invention of tradition: the Highland tradition of Scotland', in E. Hobsbawm and T. Ranger (eds), *The Invention of Tradition*. New York: Cambridge University Press. pp. 15–42.

Tuan, Y. F. (1977) *Space and Place*. Minneapolis: University of Minnesota Press.

Tunander, O. (2001) 'Swedish–German geopolitics for a new century: Rudolf Kjellén's "The state as a living organism"', *Review of International Studies*, 27(3): 451–463.

US Army (1990) Field Manual FM 100–20/AFP 3–20: *Military Operations in Low Intensity Conflict*. Washington, DC: Departments of the Army and Air Force. Retrieved 3, October 2007 from http://www.globalsecurity.org/military/library/policy/army/fm/100-20/index.html

Valentine, G. (1994) 'Toward a geography of the lesbian community', *Women and Environments*, 14(1): 8–10.

Valentine, G. (2007) 'Theorizing and research intersectionality: a challenge for feminist geography', *Professional Geographer*, 59(1): 10–21.

van der Veen, M. (2000) 'Beyond slavery and capitalism: producing class difference', in J. K. Gibson-Graham, S. Resnick and R. Wolff (eds), *Class and its Others*. Minneapolis: University of Minnesota Press. pp. 121–142.

Wade, C. (2006) Editorial: 'A historical case for the role of regional geography in geographic education', *Journal of Geography in Higher Education*, 30(2): 181–189.

Walker, R. B. J. (1993) *Inside/Outside: International Relations as Political Theory*. New York: Cambridge University Press.

Wallerstein, I. (1974) *The Modern World-system, Vol. I: Capitalist Agriculture and the Origins of the European World-economy in the Sixteenth Century*. New York and London: Academic Press.

Wallerstein, I. (2003) *The Decline of American Power*. New York: The New Press.

Wallerstein, I. (2006) 'Remembering Andre Gunder Frank', *History Workshop Journal*, 61(1): 305–306.

Ward, K. and Jonas, A. E. G. (2004) 'Competitive city-regionalism as a politics of space: a critical reinterpretation of the new regionalism', *Environment and Planning A*, 36(12): 2119–2139.

Waters, J. (2002) 'Flexible families? "Astronaut" households and the experiences of lone mothers in Vancouver, British Columbia', *Social & Cultural Geography*, 3(2): 117–134.

Wendt, A. (1999) *Social Theory of International Politics*. New York: Cambridge University Press.

West, C. (1993) *Race Matters*. Boston, MA: Beacon University Press.

Whatmore, S. (2002) *Hybrid Geographies: Natures, Cultures, Spaces*. London: Sage.

Whelan, Y. (2003) *Reinventing Modern Dublin: Streetscape, Iconography and the Politics of Identity*. Dublin: University College Dublin Press.

Whittaker, D. (2001) *The Terrorism Reader*. London: Routledge.

Wieviorka, M. (2005) 'From Marx and Braudel to Wallerstein', *Contemporary Sociology: A Journal of Reviews*, 34(1): 1–7.

Williams, M. (1996) 'Rethinking sovereignty', in E. Kofman and G. Youngs (eds), *Globalization: Theory and Practice*. London: Pinter. (pp. 109–122).

Williams, R. (1988) *Keywords: A Vocabulary of Culture and Society*, revised and expanded edn. London: Fontana Press.

Wills, J. (2006) 'The left, its crisis and rehabilitation', *Antipode*, 38(5): 907–915.

Winders, J. (2007) 'Bringing back the (B)order: post-9/11 politics of immigration, borders, and belonging in the contemporary US south', *Antipode*, 39(5): 920–942.

Wood, W. (2003) Geosecurity, in S. Cutter, D. Richardson, and T. Wilbanks (eds), *The Geographical Dimensions of Terrorism*. New York: Routledge. pp. 213–222.

Wood, A., Valler, D., Phelps, N, Raco, M. and Shirlow, P. (2005) 'Devolution and the political representation of business interests in the UK', *Political Geography*, 24: 293–315.

Woodward, R. (2005) 'From military geography to militarism's geographies: disciplinary engagements with the geographies of militarism and military activities', *Progress in Human Geography*, 29(6): 718–740.

Wright, M. (2006) *Disposable Women and Other Myths of Global Capitalism*. New York and London: Routledge.

Wright, T. C. (2007) *State Terrorism in Latin American: Chile, Argentina, and International Human Rights*. Lanham, MD: Rowman and Littlefield.

Wyne, A. (2005) 'Suicide terrorism as strategy: Case studies of Hamas and the Kurdistan Workers Party', *Strategic Insights*, 4(7). Retrieved 29 January 2008 from http://www.ccc.nps.navy.mil/si/2005/Jul/wyneJul05.asp#references

Yeung, H. (1998) 'Capital, state and space: contesting the borderless world', *Transactions of the Institute of British Geographers*, 23: 291–309.

Yeung, H. (2002) 'The limits to globalization theory: a geographic perspective on global economic change', *Economic Geography*, 78(3): 285–306.

Young, C. and Light, D. (2001) 'Place, national identity and post-socialist transformations: an introduction', *Political Geography*, 20(8): 941–955.

Young, I.M. (2000) *Inclusion and Democracy*. Oxford: Oxford University Press.

Young, R. (2001) *Postcolonialism: An Historical Introduction*. Malden, MA: Blackwell.

Zubaida, S. (2002) 'The fragments imagine the nation: the case of Iraq', *International Journal of Middle East Studies*, 34(2): 205–215.

INDEX

NOTE: Page numbers in *italic type* refer to illustrations; page numbers followed by 'n.' refer to notes.

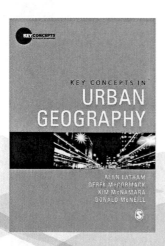

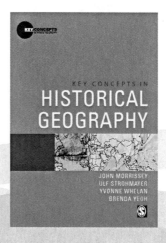

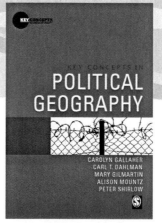